Physical Methods in Chemistry and Nano Science. Volume 6: Surface Morphology and Structure at the Nanoscale

Physical Methods in Chemistry and Nano Science. Volume 6: Surface Morphology and Structure at the Nanoscale

Editor

Andrew R. Barron

Contributors

Amir Aliyan, Andrew R. Barron, Charles Conn,
Samantha L. Donaldson, Varun Shenoy Gangoli,
Angel Adrian Garces, Daniel Garcia-Rojas, Yongji Gong,
Sravani Gullapalli, Inna Kurganskaya, Michelle LaComb, Nadia Lara,
Yen-Tien Lu, Andreas Luttge, Samuel Maguire-Boyle,
Nicole Moody, Brittany L. Oliva-Chatelain, Zhiwei Peng,
Stacy Prukop, Pavan M. V. Raja, Muqing Ren, Katherinne Requejo,
Hannah Rutledge, Richa Sethi, Macy Stavinoha, Zhengzong Sun,
Dayne Swearer, Ryan Thaner, Juan Velazquez, Zhe Wang,
Eric Wagner, Zheng Yan, Lin Yuan

MiDAS Green Innovations
2020

Cover image © 2016 by Dr_Microbe

First Printing: 2020

ISBN 978-1-8380085-9-8

9 781838 008598

MiDAS Green Innovation, Ltd
Swansea, SA1 8RD, UK

www.midasgreeninnovation.com

Dedication

To all the support staff and technicians who have assisted my research group and ensured that everything was done on time.

"The hole in the ozone layer is all right by me, makes England warmer in the summer"

Marillion (1998)

Contents

Acknowledgements

I would like to thank all the contributors to this Volume, for their interest in creating a user-friendly text for their peers.

The myriad collaborators of whom I have had the pleasure of learning from over the years are to be thanked for bringing new characterization methods to my research - you know who you are.

Last, but not least, I would like to thank my wife, Merrie, for putting up with me during the Editing of this book while 'social distancing' during the COVID-19 pandemic.

Preface

This Series intended as a survey of research techniques used in modern chemistry, materials science, and nanoscience. The topics are grouped into volumes, not be method *per se*, but with regard to the type of information that can be obtained. Thus, the Volumes are ordered as follows:

- Elemental composition.
- Physical and thermal analysis.
- Chromatography
- Chemical speciation.
- Molecular and solid-state structure.
- Surface morphology and structure at the nanoscale.
- Device performance.
- Applications of analytical methods

As a consequence of this organization methods can be found in different Volumes. For example, X-ray photoelectron spectroscopy is included under Elemental Composition (Volume 1) with regard to its use for determining the chemical composition, while it is included under Chemical Speciation (Volume 3) with regard to determining the identity of component chemical moieties.

The goal was to create simple to understand explanations of methods that allow the reader to gain the knowledge to correctly apply a technique or interpret data. As a consequence, the topics in this book have been developed in partnership with undergraduate and postgraduate students at Rice University over a 7-year period, and because of this there is some variation in depth and focus given to each topic. I make no apology for this diversity.

Chapter 1: Confocal Microscopy

Charles Conn and Andrew R. Barron

Introduction

Confocal microscopy was invented by Marvin Minsky (Figure 1.1) in 1957, and subsequently patented in 1961. Minsky was trying to study neural networks to understand how brains learn and needed a way to image these connections in their natural state (in three dimensions). He invented the confocal microscope in 1955, but its utility was not fully realized until technology could catch up. In 1973 Egger published the first recognizable cells, and the first commercial microscopes were produced in 1987.

Figure 1.1: American cognitive scientist in the field of artificial intelligence Marvin Lee Minsky (1927 - 2016).

In the 1990's confocal microscopy became near routine due to advances in laser technology, fiber optics, photodetectors, thin film dielectric coatings, computer processors, data storage, displays, and fluorophores. Today, confocal microscopy is widely used in life sciences to study cells and tissues.

The basics of fluorescence

Fluorescence is the emission of a secondary photon upon absorption of a photon of higher wavelength. Most molecules at normal temperatures are at the lowest energy state, the so-called 'ground state'. Occasionally, a molecule may absorb a photon and increase its energy to the *excited state*. From here it can very quickly transfer some of that energy to other molecules through collisions; however, if it cannot transfer enough energy it spontaneously emits a photon with a lower wavelength Figure 1.2. This is fluorescence.

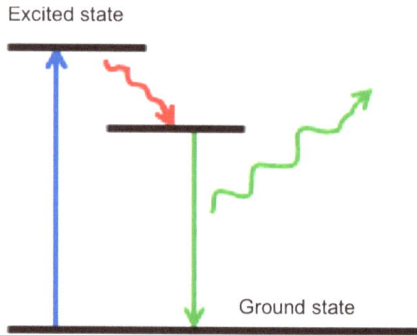

Figure 1.2: An energy diagram shows the principle of fluorescence. A molecule absorbs a high energy photon (blue) which excites the molecule to a higher energy state. The molecule then dissipates some of the extra energy via molecular collisions (red) and emits the remaining energy by emitting a photon (green) to return to the ground state.

In fluorescence microscopy, fluorescent molecules are designed to attach to specific parts of a sample, thus identifying them when imaged. Multiple fluorophores can be used to simultaneously identify different parts of a sample. There are two options when using multiple fluorophores:

- Fluorophores can be chosen that respond to different wavelengths of a multi-line laser.
- Fluorophores can be chosen that respond to the same excitation wavelength but emit at different wavelengths.

In order to increase the signal, more fluorophores can be attached to a sample. However, there is a limit, as high fluorophore concentrations result in them quenching each other, and too many fluorophores near the surface of the sample may absorb enough light to limit the light available to the rest of the sample. While the intensity of incident radiation can be increased, fluorophores may become saturated if the intensity is too high.

Photobleaching is another consideration in fluorescent microscopy. Fluorophores irreversibly fade when exposed to excitation light. This may be due to reaction of the molecules' excited state with oxygen or oxygen radicals. There has been some success in limiting photobleaching by reducing the oxygen available or by using free-radical scavengers. Some fluorophores are more robust than others, so choice of fluorophore is very important. Fluorophores today are available that emit photons with wavelengths ranging 400 - 750 nm.

How confocal microscopy is different from optical microscopy

A microscope's lenses project the sample plane onto an image plane. An image can be formed at many image planes; however, we only consider one of these planes to be the 'focal plane' (when the sample image is in focus). When a pinhole screen in placed at the image focal point, it allows in-focus light to pass while effectively blocking light from out-of-focus locations Figure 1.3. This pinhole is placed at the conjugate image plane to the focal plane, thus the name "confocal". The size of this pinhole determines the depth-of-focus; a bigger pinhole collects light from a larger volume. The pinhole can only practically be made as small as approximately the radius of the Airy disk, which is the best possible light spot from a circular aperture Figure 1.4, because beyond that more signal is blocked resulting in a decreased signal-to-noise ratio.

In optics, the Airy disk and Airy pattern are descriptions of the best focused spot of light that a perfect lens with a circular aperture can make, limited by the diffraction of light.

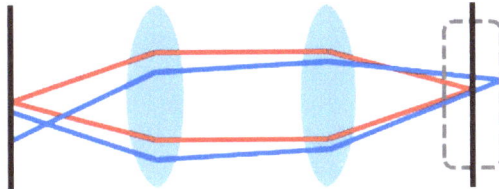

Figure 1.3: A schematic of a simplified microscope objective. Red and blue lines represent light rays refracted through the objective, indicating the focal points and corresponding image points.

To further reduce the effect of scattering due to light from other parts of the sample, the sample is only illuminated at a tiny point through the use of a pinhole in front of the light source. This greatly reduces the interference of scattered light from other parts of the sample. The combination of a pinhole in front of both the light source and detector is what makes confocal unique.

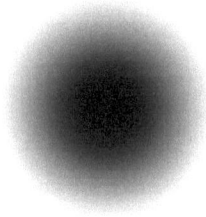

Figure 1.4: A representation of an Airy disk. An intense peak of light forms at the middle, surrounded by rings of lower intensity formed due to the diffraction of light. Adapted from Confocal Microscopy, Eric Weeks

Parts of a confocal microscope

A simple confocal microscope generally consists of a laser, pinhole aperture, dichromatic mirror, scanning mirrors, microscope objectives, a photomultiplier tube, and computing software used to reconstruct the image Figure 1.5. Because a relatively small volume of the sample is being illuminated at any given time, a very bright light source must be used to produce a detectable signal. Early confocal microscopes used zirconium arc lamps, but recent advances in laser technology have made lasers in the UV-visible and infrared more stable and affordable. A laser allows for a monochromatic (narrow wavelength range) light source that can be used to selectively excite fluorophores to emit photons of a different wavelength. Sometimes filters are used to further screen for single wavelengths.

The light passes through a dichromatic (or "dichroic") mirror Figure 1.5, which allows light with a higher wavelength (from the laser) to pass but reflects light of a lower wavelength (from the sample) to the detector. This allows the light to travel the same path through the majority of the instrument and eliminates signal due to reflection of the incident light.

The light is then reflected across a pair of mirrors or crystals, one each for the x and y directions, which enable the beam to scan across the sample (Figure 1.5). The speed of the scan is usually the limiting factor in the speed of image acquisition. Most confocal microscopes can create an image in 0.1 - 1 second. Usually the sample is raster scanned quickly in the *x*-direction and slowly in the *y* direction (like reading a paragraph left to right, Figure 1.6).

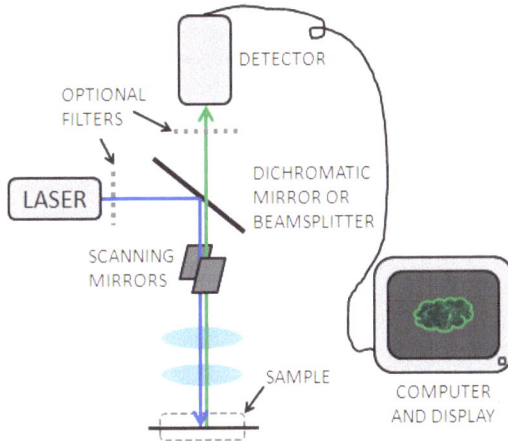

Figure 1.5: A schematic of a confocal microscope. Rays represent the path of light from source to detector.

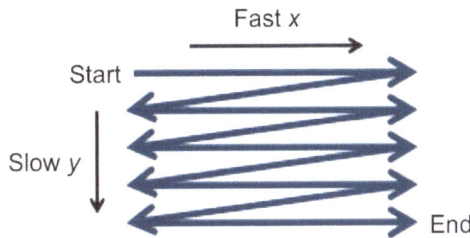

Figure 1.6: Raster scanning is usually performed quickly in the *x* direction, line-by-line. Other scanning patterns are also used, but this is most common.

The rastering is controlled by galvanometers that move the mirrors back and forth in a sawtooth motion. The disadvantage to scanning with the light beam is that the angle of light hitting the sample changes. Fortunately, this change is small. Interestingly, Minsky's original design moved the stage instead of the beam, as it was difficult to maintain alignment of the sensitive optics. Despite the obvious disadvantages of moving a bulky specimen, there are some advantages of moving the stage and keeping the optics stationary:

- The light illuminates the specimen axially everywhere circumventing optical aberrations, and
- The field of view can be made much larger by controlling the amplitude of the stage movements.

An alternative to light-reflecting mirrors is the acousto-optic deflector (AOD). The AOD allows for fast x-direction scans by creating a diffraction grating from high-frequency standing sound (pressure) waves which locally change the refractive index of a crystal. The disadvantage to AODs is that the amount of deflection depends on the wavelength, so the emission light cannot be descanned (travel back through the same path as the excitation light). The solution to this is to descan only in the y direction controlled by the slow galvanometer and collect the light in a slit instead of a pinhole. This results in reduced optical sectioning and slight distortion due to the loss of radial symmetry, but good images can still be formed. Keep in mind this is not a problem for reflected light microscopy which has the same wavelength for incident and reflected light!

Another alternative is the Nipkow disk, which has a spiral array of pinholes that create the simultaneous sampling of many points in the sample. A single rotation covers the entire specimen several times over (at 40 revolutions per second, that's over 600 frames per second). This allows descanning, but only about 1% of the excitation light passes through. This is okay for reflected light microscopy, but the signal is relatively weak and signal-to-noise ratio is low. The pinholes could be made bigger to increase light transmission but then the optical sectioning is less effective (remember depth of field is dependent on the diameter of the pinhole) and xy resolution is poorer. Highly responsive, efficient fluorophores are needed with this method.

Returning to the confocal microscope (Figure 1.5), light then passes through the objective which acts as a well-corrected condenser and objective combination. The illuminated fluorophores fluoresce, and emitted light travels up the objective back to the dichromatic mirror. This is known as epifluorescence when the incident light has the same path as detected light. Since the emitted light now has a lower wavelength than the incident, it cannot pass through the dichromatic mirror and is reflected to the detector. When using reflected light, a beamsplitter is used instead of a dichromatic mirror. Fluorescence microscopy when used properly can be more sensitive than reflected light microscopy.

Though the signal's position is well-defined according to the position of the *xy* mirrors, the signal from fluorescence is relatively weak after passing through the pinhole, so a photomultiplier tube is used to detect emitted photons. Detecting all photons without regard to spatial position increases the signal, and the photomultiplier tube further increases the detection signal by propagating an electron cascade resulting from the photoelectric effect (incident photons kicking off electrons). The resulting signal is an analog electrical signal with continuously varying voltage that corresponds to the emission intensity. This is periodically sampled by an analog-to-digital converter.

It is important to understand that the image is a reconstruction of many points sampled across the specimen. At any given time, the microscope is only looking at a tiny point, and no complete image exists that can be viewed at an instantaneous point in time. Software is used to recombine these points to form an image plane and combine image planes to form a 3-D representation of the sample volume.

Two-photon microscopy

Two-photon microscopy is a technique whereby two beams of lower intensity are directed to intersect at the focal point (Figure 1.7). Two photons can excite a fluorophore if they hit it at the same time, but alone they do not have enough energy to excite any molecules. The probability of two photons hitting a fluorophore at nearly the exact same time (less than 10^{-16}) is very low, but more likely at the focal point. This creates a bright point of light in the sample without the usual cone of light above and below the focal plane, since there are almost no excitations away from the focal point.

To increase the chance of absorption, an ultra-fast pulsed laser is used to create quick, intense light pulses. Since the hourglass shape is replaced by a point source, the pinhole near the detector (used to reduce the signal from light originating from outside the focal plane) can be eliminated. This also increases the signal-to-noise ratio (here is very little noise now that the light source is so focused, but the signal is also small). These lasers have lower average incident power than normal lasers, which helps reduce damage to the surrounding specimen. This technique can image deeper into the specimen (~400 μm), but these lasers are still very expensive, difficult to set up, require a stronger power supply, intensive cooling, and must be aligned in the same optical table because pulses can be distorted in optical fibers.

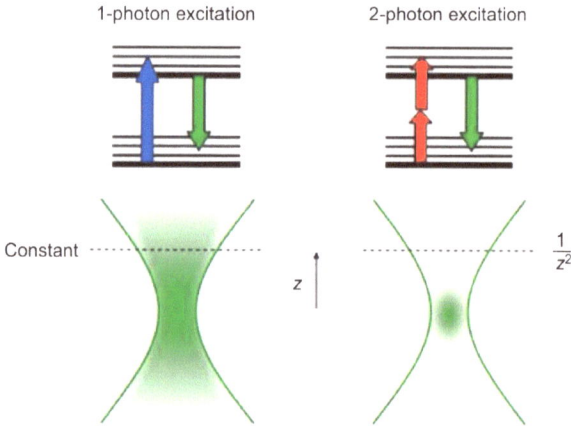

Figure 1.7: Schematic representation of the difference between single photon and two photon microscopy. Copyright: J. Mertz, Boston University.

Microparticle characterization

Confocal microscopy is very useful for determining the relative positions of particles in three dimensions (Figure 1.8). Software allows measurement of distances in the 3D reconstructions so that information about spacing can be ascertained (such as packing density, porosity, long range order or alignment, etc.).

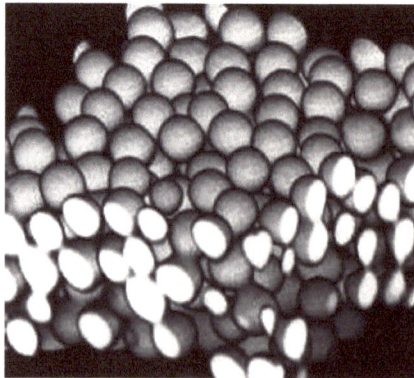

Figure 1.8: A reconstruction of a colloidal suspension of poly(methyl methacrylate) (PMMA, Figure 1.9) microparticles approximately 2 microns in diameter. Adapted from Confocal Microscopy of Colloids, Eric Weeks.

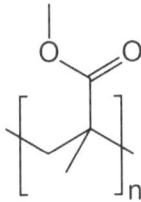

Figure 1.9: Structure of poly(methyl methacrylate) (PMMA).

If imaging in fluorescence mode, remember that the signal will only represent the locations of the individual fluorophores. There is no guarantee fluorophores will completely attach to the structures of interest or that there will not be stray fluorophores away from those structures. For microparticles it is often possible to attach the fluorophores to the shell of the particle, creating hollow spheres of fluorophores. It is possible to tell if a sample sphere is hollow or solid, but it would depend on the transparency of the material.

Dispersions of microparticles have been used to study nucleation and crystal growth, since colloids are much larger than atoms and can be imaged in real-time. Crystalline regions are determined from the order of spheres arranged in a lattice, and regions can be distinguished from one another by noting lattice defects.

Self-assembly is another application where time-dependent, 3-D studies can help elucidate the assembly process and determine the position of various structures or materials. Because confocal is popular for biological specimens, the position of nanoparticles such as quantum dots in a cell or tissue can be observed. This can be useful for determining toxicity, drug-delivery effectiveness, diffusion limitations, etc.

A summary of confocal microscopy's strengths and weaknesses

Strengths

- Less haze, better contrast than ordinary microscopes.
- 3-D capability.
- Illuminates a small volume.
- Excludes most of the light from the sample not in the focal plane.

- Depth of field may be adjusted with pinhole size.
- Has both reflected light and fluorescence modes.
- Can image living cells and tissues.
- Fluorescence microscopy can identify several different structures simultaneously.
- Accommodates samples with thickness up to 100 μm.
- Can use with two-photon microscopy.
- Allows for optical sectioning (no artifacts from physical sectioning) 0.5 - 1.5 μm.

Weakness

- Images are scanned slowly (one complete image every 0.1-1 second).
- Must raster scan sample, no complete image exists at any given time.
- There is an inherent resolution limit because of diffraction (based on numerical aperture, ~200 nm).
- Sample should be relatively transparent for good signal.
- High fluorescence concentrations can quench the fluorescent signal.
- Fluorophores irreversibly photobleach.
- Lasers are expensive.
- Angle of incident light changes slightly, introducing slight distortion.

Bibliography

P. Davidovits and M. D. Egger, Photomicrography of corneal endothelial cells in vivo. *Nature*, 1973, **244**, 366.

A. Hibbs, *Confocal Microscopy for Biologists*, Twayne Publishers, Boston (2004).

M. Minsky, Memoir on inventing the confocal scanning microscope. *Scanning*, 1988, **10**, 128.

J. Pawley, *Handbook of Biological Confocal Microscopy*, Twayne Publishers, Boston (2006).

V. Prasad, D. Semwogerere, and E.R. Weeks, Confocal microscopy of colloids. *J. Phys. Condens. Matter*, 2007, **19**, 113102.

D. Semwogerere and E. R. Weeks, *Encyclopedia of Biomaterials and Biomedical Engineering Confocal Microscopy*, Taylor Francis, Abingdon-on-Thames (2005).

C. Sheppard, *Confocal Laser Scanning Microscopy*, Twayne Publishers, Boston (1997).

Chapter 2: Vertical Scanning Interferometry

Inna Kurganskaya, Andreas Luttge and Andrew R. Barron

Introduction

The processes which occur at the surfaces of crystals depend on many external and internal factors such as crystal structure and composition, conditions of a medium where the crystal surface exists and others. The appearance of a crystal surface is the result of complexity of interactions between the crystal surface and the environment. The mechanisms of surface processes such as dissolution or growth are studied by the physical chemistry of surfaces. There are a lot of computational techniques which allows us to predict the changing of surface morphology of different minerals which are influenced by different conditions such as temperature, pressure, pH and chemical composition of solution reacting with the surface. For example, Monte Carlo method is widely used to simulate the dissolution or growth of crystals. However, the theoretical models of surface processes need to be verified by natural observations. We can extract a lot of useful information about the surface processes through studying the changing of crystal surface structure under influence of environmental conditions. The changes in surface structure can be studied through the observation of crystal surface topography. The topography can be directly observed macroscopically or by using microscopic techniques. Microscopic observation allows us to study even very small changes and estimate the rate of processes by observing changing the crystal surface topography in time.

Much laboratory worked under the reconstruction of surface changes and interpretation of dissolution and precipitation kinetics of crystals. Invention of AFM made possible to monitor changes of surface structure during dissolution or growth. However, to detect and quantify the results of dissolution processes or growth it is necessary to determine surface area changes over a significantly larger field of view than AFM can provide. More recently, vertical scanning interferometry (VSI) has been developed as new tool to distinguish and trace the reactive parts of crystal surfaces. VSI and AFM are complementary techniques and practically well suited to detect surface changes.

VSI technique provides a method for quantification of surface topography at the angstrom to nanometer level. Time-dependent VSI measurements can be used to study the surface-normal retreat across crystal and other solid surfaces during dissolution process. Therefore, VSI can be used to directly and non-directly measure mineral dissolution rates with high precision. Analogically, VSI can be used to study kinetics of crystal growth.

Physical principles of optical interferometry

Optical interferometry allows us to make extremely accurate measurements and has been used as a laboratory technique for almost a hundred years. Thomas Young observed interference of light and measured the wavelength of light in an experiment, performed around 1801. This experiment gave an evidence of Young's arguments for the wave model for light. The discovery of interference gave a basis to development of interferomertry techniques widely successfully used as in microscopic investigations, as in astronomic investigations.

The physical principles of optical interferometry exploit the wave properties of light. Light can be thought as electromagnetic wave propagating through space. If we assume that we are dealing with a linearly polarized wave propagating in a vacuum in z direction, electric field E can be represented by a sinusoidal function of distance and time.

$$E(x,y,z,t) = a\cos[2\pi(vt - z/\lambda)]$$

where a is the amplitude of the light wave, v is the frequency, and λ is its wavelength. The term within the square brackets is called the phase of the wave. Let's rewrite this equation in more compact form,

$$E(x,y,z,t) = a\cos[\omega t - kz]$$

where

$$\omega = 2\pi v$$

is the circular frequency, and

$$k = 2\pi/\lambda$$

is the propagation constant. Let's also transform this second equation into a complex exponential form,

$$E(x,y,z,t) = \text{Re}\{a\exp(i\phi)\exp(i\omega t)\} = \text{Re}\{A\exp(i\omega t)\}$$

where

$$\phi = 2\pi z/\lambda$$

and

$$A = \exp(-i\phi)$$

is known as the complex amplitude. If n is a refractive index of a medium where the light propagates, the light wave traverses a distance d in such a medium. The equivalent optical path in this case is

$$P = n \times d$$

When two light waves are superposed, the result intensity at any point depends on whether reinforce or cancel each other (Figure 2.1). This is well known phenomenon of interference. We will assume that two waves are propagating in the same direction and are polarized with their field vectors in the same plane. We will also assume that they have the same frequency. The complex amplitude at any point in the interference pattern is then the sum of the complex amplitudes of the two waves, so that we can write,

$$A = A_1 + A_2$$

where

$$A_1 = a_1\exp(-i\phi_1)$$

and

$$A_2 = a_2\exp(-i\phi_2)$$

are the complex amplitudes of two waves. The resultant intensity is, therefore,

$$I = |A|^2 = I_1 + I_2 + 2(I_1I_2)^{1/2}\cos\Delta\phi$$

where I_1 and I_2 are the intensities of two waves acting separately, and

$$\Delta\phi = \phi_1 - \phi_2$$

is the phase difference between them. If the two waves are derived from a common source, the phase difference corresponds to an optical path difference,

$$\Delta p = (\lambda/2\pi)\Delta\phi$$

Two waves in phase

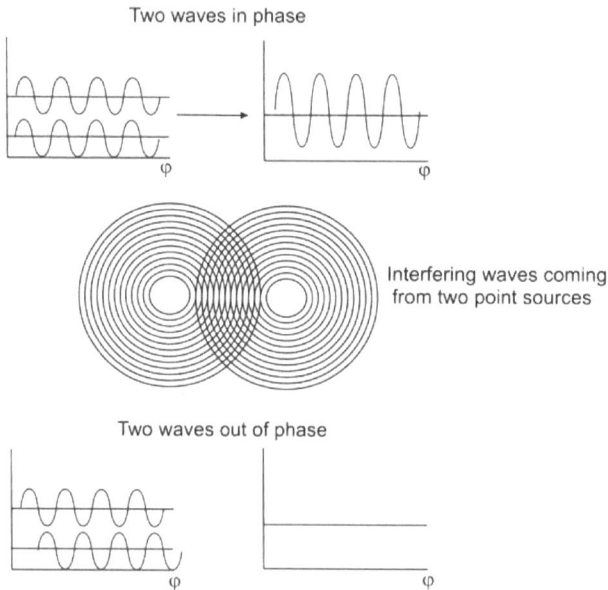

Interfering waves coming from two point sources

Two waves out of phase

Figure 2.1: The scheme of interferometric wave interaction when two waves interact with each other, the amplitude of resulting wave will increase or decrease. The value of this amplitude depends on phase difference between two original waves.

If $\Delta\phi$, the phase difference between the beams, varies linearly across the field of view, the intensity varies cosinusoidally, giving rise to alternating light and dark bands or fringes (Figure 2.1). The intensity in an interference pattern has its maximum value

$$I_{max} = I_1 + I_2 + 2(I_1 I_2)^{1/2}$$

when

$$\Delta\phi = 2m\pi$$

, where m is an integer and its minimum value

$$I_{min} = I_1 + I_2 - 2(I_1 I_2)^{1/2}$$

when

$$\Delta\phi = (2m + 1)\pi$$

The principle of interferometry is widely used to develop many types of interferometric set ups. One of the earliest set ups is Michelson interferometry. The idea of this interferometry is quite simple: interference fringes are produced by splitting a beam of monochromatic light so that one beam strikes a fixed mirror and the other a movable mirror. An interference pattern results when the reflected beams are brought back together. The Michelson interferometric scheme is shown in Figure 2.2.

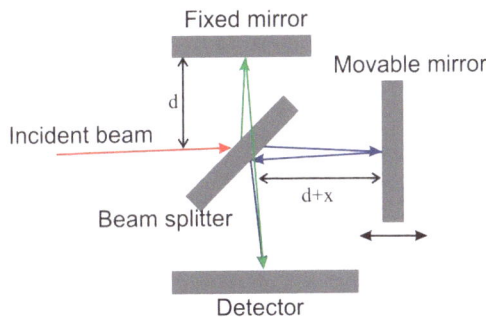

Figure 2.2: Schematic representation of a Michelson interferometry set-up.

The difference of path lengths between two beams is 2x because beams traverse the designated distances twice. The interference occurs when the path difference is equal to integer numbers of wavelengths,

$$\Delta p = 2x = m\lambda, m = 0, \pm 1, \pm 2...$$

Modern interferometric systems are more complicated. Using special phase-measurement techniques they capable to perform much more accurate height measurements than can be obtained just by directly looking at the interference fringes and measuring how they depart from being straight and equally spaced. Typically, interferometric system consists of lights source, beamsplitter, objective system, system of registration of signals and transformation into digital format and computer which process data. Vertical scanning interferometry contains all these parts. Figure 2.3 shows a configuration of VSI interferometric system.

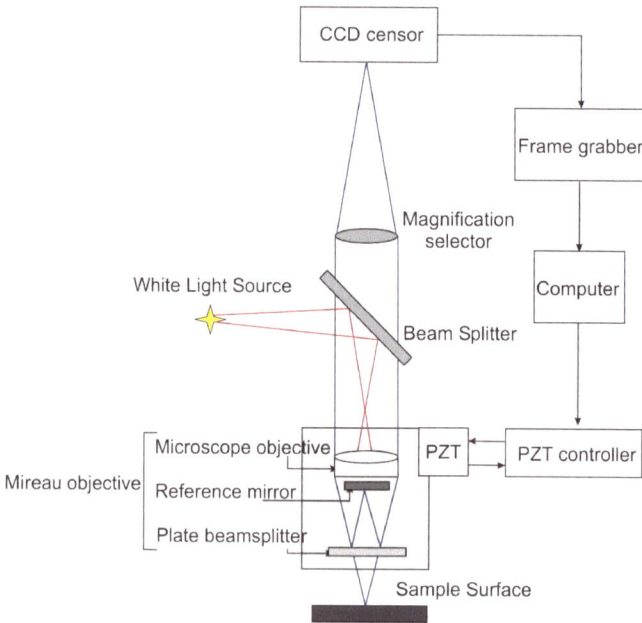

Figure 2.3: Schematic representation of the Vertical scanning interferometry (VSI) system.

Many of modern interferometric systems use Mirau objective in their constructions. Mireau objective is based on a Michelson interferometer. This objective consists of a lens, a reference mirror and a beamsplitter. The idea of getting interfering beams is simple: two beams (red lines) travel along the optical axis. Then they are reflected from the reference surface and the sample surface respectively (blue lines). After this these beams are recombined to interfere with each other. An illumination or light source system is used to direct light onto a sample surface through a cube beam splitter and the Mireau objective. The sample surface within the field of view of the objective is uniformly illuminated by those beams with different incidence angles. Any point on the sample surface can reflect those incident beams in the form of divergent cone. Similarly, the point on the reference symmetrical with that on the sample surface also reflects those illuminated beams in the same form.

The Mireau objective directs the beams reflected of the reference and the sample surface onto a CCD (charge-coupled device) sensor through a tube lens. The CCD sensor is an analog shift register that enables the transportation of analog signals (electric charges) through successive stages (capacitors), controlled by a clock signal. The resulting interference fringe pattern is detected by CCD sensor and the corresponding signal is digitized by a frame grabber for further processing with a computer.

The distance between a minimum and a maximum of the interferogram produced by two beams reflected from the reference and sample surface is known. That is, exactly half the wavelength of the light source. Therefore, with a simple interferogram the vertical resolution of the technique would be also limited to $\lambda/2$. If we will use a laser light as a light source with a wavelength of 300 nm the resolution would be only 150 nm. This resolution is not good enough for a detailed near-atomic scale investigation of crystal surfaces. Fortunately, the vertical resolution of the technique can be improved significantly by moving either the reference or the sample by a fraction of the wavelength of the light. In this way, several interferograms are produced. Then they are all overlaid, and their phase shifts compared by the computer software (Figure 2.4). This method is widely known as phase shift interferometry (PSI).

Multiple interferograms
of the surface

Sample surface

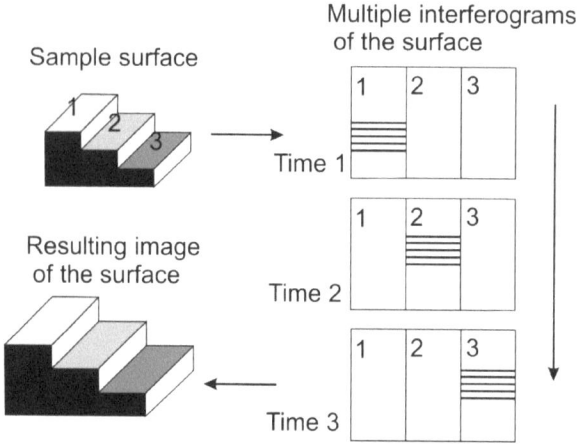

Time 1

Resulting image
of the surface

Time 2

Time 3

Figure 2.4: Sketch illustrating phase-shift technology. The sample is continuously moved along the vertical axes in order to scan surface topography. All interferograms are automatically overlaid using computer software.

Most optical testing interferometers now use phase-shifting techniques not only because of high resolution but also because phase-shifting is a high accuracy rapid way of getting the interferogram information into the computer. Also, usage of this technique makes the inherent noise in the data taking process very low. As the result in a good environment angstrom or sub-angstrom surface height measurements can be performed. As it was said above, in phase-shifting interferometry the phase difference between the interfering beams is changed at a constant rate as the detector is read out. Once the phase is determined across the interference field, the corresponding height distribution on the sample surface can be determined. The phase distribution $\varphi(x, y)$ is recorded by using the CCD camera.

Let's assign $A(x, y)$, $B(x, y)$, $C(x, y)$ and $D(x, y)$ to the resulting interference light intensities which are corresponded to phase-shifting steps of 0, $\pi/2$, π and $3\pi/2$. These intensities can be obtained by moving the reference mirror through displacements of $\lambda/8$, $\lambda/4$ and $3\lambda/8$, respectively. The equations for the resulting intensities would be:

$$A(x,y) = I_1(x,y) + I_2(x,y)\cos\alpha(x,y)$$

$$B(x,y) = I_1(x,y) - I_2(x,y)\sin\alpha(x,y)$$

$$C(x,y) = I_1(x,y) - I_2(x,y)\cos\alpha(x,y)$$

$$D(x,y) = I_1(x,y) + I_2(x,y)\sin\alpha(x,y)$$

where $I_1(x,y)$ and $I_2(x,y)$ are two overlapping beams from two symmetric points on the test surface and the reference respectively. Solving these equations the phase map $\phi(x,y)$ of a sample surface will be given by the relation:

$$\phi(x,y) = \frac{B(x,y) - D(x,y)}{A(x,y) - C(x,y)}$$

Once the phaseis determined across the interference field pixel by pixel on a two-dimensional CCD array, the local height distribution/contour, $h(x,y)$, on the test surface is given by

$$h(x,y) = \frac{\lambda}{4\pi}\,\phi(x,y)$$

Normally the resulted fringe can be in the form of a linear fringe pattern by adjusting the relative position between the reference mirror and sample surfaces. Hence any distorted interference fringe would indicate a local profile/contour of the test surface.

It is important to note that the Mireau objective is mounted on a capacitive closed-loop controlled PZT (piezoelectric actuator) as to enable phase shifting to be accurately implemented. The PZT is based on piezoelectric effect referred to the electric potential generated by applying pressure to piezoelectric material. This type of materials is used to convert electrical energy to mechanical energy and vice-versa. The precise motion that results when an electric potential is applied to a piezoelectric material has an importance for nanopositioning. Actuators using the piezo effect have been commercially available for 35 years and in that time have transformed the world of precision positioning and motion control.

Vertical scanning interferometer also has another name; white-light interferometry (WLI) because of using the white light as a source of light. With this type of source, a separate fringe system is produced for each wavelength, and the resultant intensity at any point of examined surface is obtained by summing these individual patterns. Due to the broad bandwidth of the source the coherent length L of the source is short:

$$L = \frac{\lambda^2}{n\Delta\lambda}$$

where λ is the center wavelength, n is the refractive index of the medium, $\Delta\lambda$ is the spectral width of the source. In this way good contrast fringes can be obtained only when the lengths of interfering beams pathways are closed to each other. If we will vary the length of a pathway of a beam reflected from sample, the height of a sample can be determined by looking at the position for which a fringe contrast is a maximum. In this case interference pattern exist only over a very shallow depth of the surface. When we vary a pathway of sample-reflected beam we also move the sample in a vertical direction in order to get the phase at which maximum intensity of fringes will be achieved. This phase will be converted in height of a point at the sample surface.

The combination of phase shift technology with white-light source provides a very powerful tool to measure the topography of quite rough surfaces with the amplitude in heights about and the precision up to 1 - 2 nm. Through a developed software package for quantitatively evaluating the resulting inter-ferogram, the proposed system can retrieve the surface profile and topography of the sample objects (Figure 2.5).

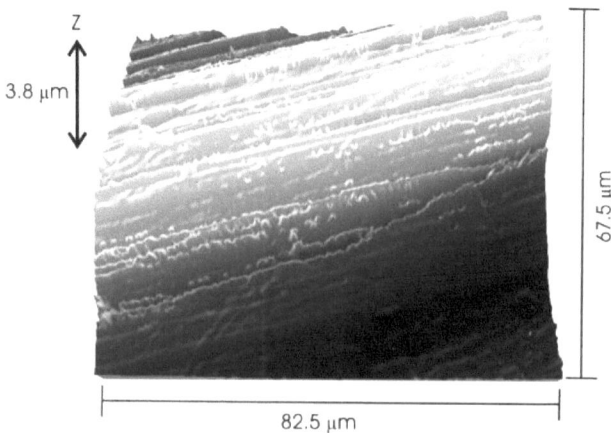

Figure 2.5: Example of muscovite surface topography, obtained by using VSI-50× objective.

A comparison of common methods to determine surface topography: SEM, AFM and VSI

Except the interferometric methods described above, there are a several other microscopic techniques for studying crystal surface topography. The most common are scanning electron microscopy (SEM) and atomic force microscopy (AFM). All these techniques are used to obtain information about the surface structure. However, they differ from each other by the physical principles on which they based.

Scanning electron microscopy

SEM allows us to obtain images of surface topography with the resolution much higher than the conventional light microscopes do. Also it is able to provide information about other surface characteristics such as chemical composition, electrical conductivity etc. (Figure 2.6). All types of data are generated by the reflecting of accelerated electron beams from the sample surface. When electrons strike the sample surface, they lose their energy by repeated random scattering and adsorption within an outer layer into the depth varying from 100 nm to 5 microns.

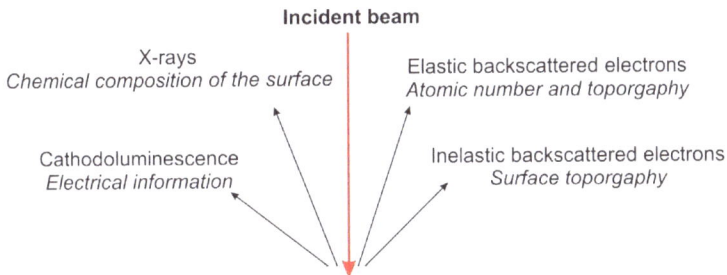

Incident beam

X-rays
Chemical composition of the surface

Elastic backscattered electrons
Atomic number and toporgaphy

Cathodoluminescence
Electrical information

Inelastic backscattered electrons
Surface toporgaphy

Figure 2.6: Scheme of electron beam-sample interaction at SEM analysis.

The thickness of this outer layer also knows as interactive layer depends on energy of electrons in the beam, composition and density of a sample. Result of the interaction between electron beam and the surface provides several types of signals. The main type is secondary or inelastic scattered electrons. They are produced as a result of interaction between the beam of electrons and weakly bound electrons in the conduction band of the sample. Secondary electrons are ejected from the k-orbitals of atoms within the surface layer of thickness about a few nanometers. This is because secondary electrons are low energy electrons (<50 eV), so only those formed within the first few nanometers of the sample surface have enough energy to escape and be detected. Secondary backscattered electrons provide the most common signal to investigate surface topography with lateral resolution up to 0.4 - 0.7 nm.

High energy beam electrons are elastic scattered back from the surface. This type of signal gives information about chemical composition of the surface because the energy of backscattered electrons depends on the weight of atoms within the interaction layer. Also, this type of electrons can form secondary electrons and escape from the surface or travel father into the sample than the secondary. The SEM image formed is the result of the intensity of the secondary electron emission from the sample at each x,y data point during the scanning of the surface.

Atomic force microscopy

AFM is a very popular tool to study surface dissolution. AFM set up consists of scanning a sharp tip on the end of a flexible cantilever which moves across a sample surface. The tips typically have an end radius of 2 to 20 nm, depending on tip type. When the tip touches the surface the forces of these interactions leads to deflection of a cantilever. The interaction between tip and sample surface involve mechanical contact forces, van der Waals forces, capillary forces, chemical bonding, electrostatic forces, magnetic forces etc. The deflection of a cantilever is usually measured by reflecting a laser beam off the back of the cantilever into a split photodiode detector. A schematic drawing of AFM can be seen in Figure 2.7. The two most commonly used modes of operation are contact mode AFM and tapping mode AFM, which are conducted in air or liquid environments.

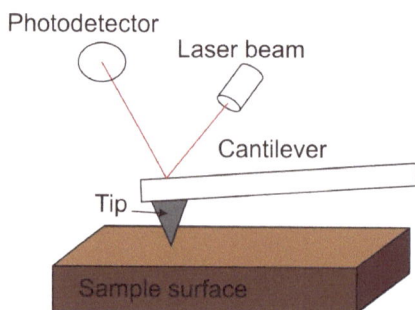

Figure 2.7: Schematic drawing of an AFM apparatus.

Working under the contact mode AFM scans the sample while monitoring the change in cantilever deflection with the split photodiode detector. Loop maintains a constant cantilever reflection by vertically moving the scanner to get a constant signal. The distance which the scanner goes by moving vertically at each x,y data point is stored by the computer to form the topographic image of the sample surface. Working under the tapping mode AFM oscillates the cantilever at its resonance frequency (typically~300 kHz) and lightly "taps" the tip on the surface during scanning. The electrostatic forces increase when tip gets close to the sample surface, therefore the amplitude of the oscillation decreases. The laser deflection method is used to detect the amplitude of cantilever oscillation. Similar to the contact mode, feedback loop maintains a constant oscillation amplitude by moving the scanner vertically at every x,y data point. Recording this movement forms the topographical image. The advantage of tapping mode over contact mode is that it eliminates the lateral, shear forces present in contact mode. This enables tapping mode to image soft, fragile, and adhesive surfaces without damaging them while work under contact mode allows the damage to occur.

Comparison of techniques

All techniques described above are widely used in studying of surface nano- and micromorphology. However, each method has its own limitations and the proper choice of analytical technique depends on features of analyzed surface and primary goals of research.

All these techniques are capable to obtain an image of a sample surface with quite good resolution. The lateral resolution of VSI is much less, then for other techniques: 150 nm for VSI and 0.5 nm for AFM and SEM. Vertical resolution of AFM (0.5 Å) is better than for VSI (1 - 2 nm), however VSI is capable to measure a high vertical range of heights (1 mm) which makes possible to study even very rough surfaces. On the contrary, AFM allows us to measure only quite smooth surfaces because of its relatively small vertical scan range (7 μm). SEM has less resolution, than AFM because it requires coating of a conductive material with the thickness within several nm.

The significant advantage of VSI is that it can provide a large field of view (845 × 630 μm for 10× objective) of tested surfaces. Recent studies of surface roughness characteristics showed that the surface roughness parameters increase with the increasing field of view until a critical size of 250,000 μm is reached. This value is larger than the maximum field of view produced by

AFM (100×100 μm) but can be easily obtained by VSI. SEM is also capable to produce images with large field of view. However, SEM is able to provide only 2D images from one scan while AFM and VSI let us to obtain 3D images. It makes quantitative analysis of surface topography more complicated, for example, topography of membranes is studied by cross section and top view images.

	VSI	AFM	SEM
Lateral resolution	0.5-1.2 μm	0.5 nm	0.5-1 nm
Vertical resolution	2 nm	0.5 Å	Only 2D images
Field of view	845×630 μm (10× objective)	100×100 μm	1-2 mm
Vertical range of scan	1 mm	10 μm	-
Preparation of a sample	-	-	Required coating of a conducted material
Required environment	Air	Air, liquid	Vacuum

Table 2.1: A comparison of VSI sample and resolution with AFM and SEM.

The experimental studying of surface processes using microscopic techniques

The limitations of each technique described above are critically important to choose appropriate technique for studying surface processes. Let's explore application of these techniques to study dissolution of crystals.

When crystalline matter dissolves the changes of the crystal surface topography can be observed by using microscopic techniques. If we will apply an unreactive mask (silicon for example) on crystal surface and place a crystalline sample into the experiment reactor then we get two types of surfaces: dissolving and remaining the same or unreacted. After some period of time the crystal surface starts to dissolve and change its z-level. In order to study these changes *ex situ* we can pull out a sample from the reaction cell then remove a mask and measure the average height difference Δh between the unreacted and dissolved areas. The average heights of dissolved and unreacted areas are obtained through digital processing of data obtained by microscopes. The velocity of normal surface retreat vSNR during the time interval Δt is defined as

$$v_{SNR} = \frac{\Delta h}{\Delta t}$$

Dividing this velocity by the molar volume $V(cm^3/mol)$ gives a global dissolution rate in the familiar units of moles per unit area per unit time:

$$R = \frac{v_{SNR}}{V}$$

This method allows us to obtain experimental values of dissolution rates just by precise measuring of average surface heights. Moreover, using this method we can measure local dissolution rates at etch pits by monitoring changes in the volume and density of etch pits across the surface over time. VSI technique is capable to perform these measurements because of large vertical range of scanning. In order to get precise values of rates which are not depend on observing place of crystal surface we need to measure enough large areas. VSI technique provides data from areas which are large enough to study surfaces with heterogeneous dissolution dynamics and obtain average dissolution rates. Therefore, VSI makes possible to measure rates of normal surface retreat during the dissolution and observe formation, growth and distribution of etch pits on the surface.

However, if the mechanism of dissolution is controlled by dynamics of atomic steps and kink sites within a smooth atomic surface area, the observation of the dissolution process needs to use a more precise technique. AFM is capable to provide information about changes in step morphology *in situ* when the dissolution occurs. For example, immediate response of the dissolved surface to the changing of environmental conditions (concentrations of ions in the solution, pH etc.) can be studied by using AFM.

SEM is also used to examine micro and nanotexture of solid surfaces and study dissolution processes. This method allows us to observe large areas of crystal surface with high resolution which makes possible to measure a high variety of surfaces. The significant disadvantage of this method is the requirement to cover examine sample by conductive substance which limits the resolution of SEM. The other disadvantage of SEM is that the analysis is conducted in vacuum. Recent technique, environmental SEM or ESEM overcomes these requirements and makes possible even examine liquids and biological materials. The third disadvantage of this technique is that it produces only 2D images. This creates some difficulties to measure Δh within the dissolving area. One of advantages of this technique is that it is able to measure not only surface topography but also chemical composition and other surface characteristics of the surface. This fact is used to monitor changing in chemical composition during the dissolution.

Bibliography

K. Babic-Samardzija, C. Lupu, N. Hackerman, A. R. Barron, and A. Luttge, Inhibitive properties and surface morphology of a group of heterocyclic diazoles as inhibitors for acidic iron corrosion. *Langmuir*, 2005, **21**, 12187

K. J. Davis and A. Luttge, Quantifying the relationship between microbial attachment and mineral surface dynamics using vertical scanning interferometry (VSI). *Am. J. Sci.*, 2005, **305**, 727.

C. Fischer and A. Luttge, Converged surface roughness parameters - A new tool to quantify rock surface morphology and reactivity alteration. *Am. J. Sci.*, 2007, **307**, 955.

P. Hariharan. *Optical interferometry*, 2nd edn., Academic Press, Cambridge MA (2003).

D. Kaczmarek, Investigation of surface topography using a multidetector system in a SEM. *Vacuum*, 2001, **62**, 303.

A. C. Lasaga, *Kinetic Theory in the Earth Sciences*, Princeton Univ. Press, Princeton, NJ (1998).

A. C. Lasaga and A. Luttge, Mineralogical approaches to fundamental crystal dissolution kinetics. *Am. Mineral.*, 2004, **89**, 527.

C. Lupu, R. S. Arvidson, A. Lüttge, and A. R. Barron, Phosphonate mediated surface reaction and reorganization: implications for the mechanism controlling cement hydration inhibition. *Chem. Commun.*, 2005, 2354.

A. Luttge and R. S. Arvidson, in *Kinetics of water-rock interaction*, Ed. S. Brantley, J. Kubicki, and A. White, Springer, New York, NY (2007).

A. Luttge, E. V. Bolton, and A. C. Lasaga, An interferometric study of the dissolution kinetics of anorthite; the role of reactive surface area. *Am. J. Sci.*, 1999, **299**, 652.

A. Luttge and P. G. Conrad, Direct observation of microbial inhibition of calcite dissolution. *Appl. Environ. Microbiol.*, 2004, **70**, 1627.

T. C. Vaimakis, E. D. Economou, and C. C. Trapalis, Calorimetric study of dissolution kinetics of phosphorite in diluted acetic acid. *J. Therm. Anal. Cal.*, 2008, **92**, 783.

S. H. Wang and Tay, Application of an optical interferometer for measuring the surface contour of micro-components. *Meas. Sci. Technol.*, 2006, **17**, 617.

Y. Wyart, G. Georges, C. Deumie, C. Amra, and P. Moulina, Membrane characterization by microscopic methods: multiscale structure. *J. Membrane Sci.*, 2008, **315**, 82.

L. Zhang and A. Luttge, Al,Si order in albite and its effect on albite dissolution processes: A Monte Carlo study. *Am. Mineral.*, 2007, **92**, 1316.

Chapter 3: Dual Polarization Interferometry

Macy Stavinoha and Andrew R. Barron

Introduction

As research interests begin to focus on progressively smaller dimensions, the need for nanoscale characterization techniques has seen a steep rise in demand. In addition, the wide scope of nanotechnology across all fields of science has perpetuated the application of characterization techniques to a multitude of disciplines. Dual polarization interferometry (DPI) is an example of a technique developed to solve a specific problem but was expanded and utilized to characterize fields ranging surface science, protein studies, and crystallography. With a simple optical instrument, DPI can perform label-free sensing of refractive index and layer thickness in real time, which provides vital information about a system on the nanoscale, including the elucidation of structure-function relationships.

History

DPI was conceived in 1996 by Dr. Neville Freeman and Dr. Graham Cross when they recognized a need to measure refractive index and adlayer thickness simultaneously in protein membranes to gain a true understanding of the dynamics of the system. They patented the technique in 1998, and the instrument was commercialized by Farfield Group Ltd. in 2000.

Freeman and Cross unveiled the first full publication describing the technique in 2003, where they chose to measure well-known protein systems and compare their data to X-ray crystallography and neutron reflection data. The first system they studied was sulpho-NHS-LC-biotin coated with streptavidin and a biotinylated peptide capture antibody, and the second system was BS^3 coated with anti-HSA. Molecular structures are shown in Figure 3.1. Their results showed good agreement with known layer thicknesses, and the method showed clear advantages over neutron reflection and surface plasmon resonance. A schematic and picture of the instrument used by Freeman and Cross in this publication is shown in Figure 3.2 and Figure 3.3, respectively.

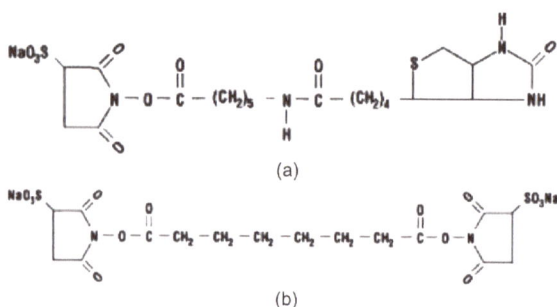

(a)

(b)

Figure 3.2: Molecular structures of (a) sulpho-NHS-LC-biotin and (b) *bis*-(sulphosuccinimydyl) suber- ate (BS³). Adapted from G. H. Cross, A. A. Reeves, S. Brand, J. F. Popplewell, L. L. Peel, M. J. Swann, and N. J. Freeman, A new quantitative optical biosensor for protein characterisation. *Biosens. Bioelectron.*, 2003, 19, 383. Copyright: Biosensors & Bioelectronics (2003).

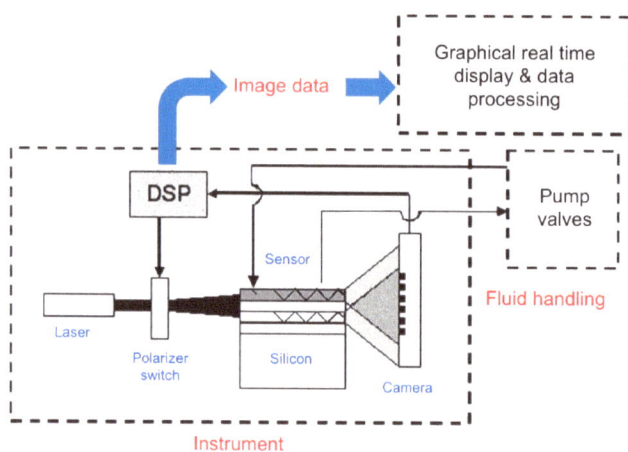

Figure 3.3: The first DPI schematic and instrument. Adapted from G. H. Cross, A. A. Reeves, S. Brand, J. F. Popplewell, L. L. Peel, M. J. Swann, and N. J. Freeman, A new quantitative optical biosensor for protein characterisation. *Biosens. Bioelectron.*, 2003, 19, 383. Copyright: Biosensors & Bioelectronics (2003).

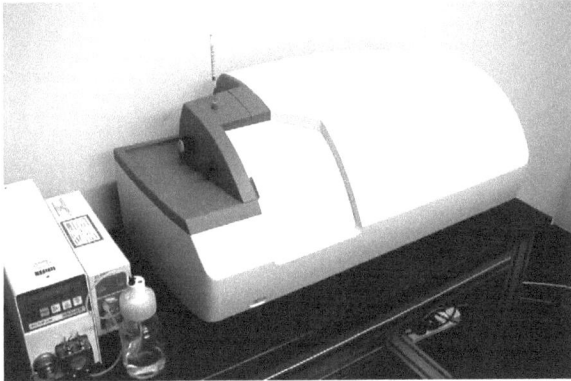

Figure 3.3: Picture of the DPI instrument used by Freeman and Cross.

Instrumentation

Theory

The optical power of DPS comes from the ability to measure two different interference fringe patterns simultaneously in real time. Phase changes in these fringe patterns result from changes in refractive index and layer thickness that can be detected by the waveguide interferometer and resolving these interference patterns provides refractive index and layer thickness values.

Optics

A representation of the interferometer is shown in Figure 3.4. The interferometer is composed of a simplified slab waveguide, which guides a wave of light in one transverse direction without scattering. A broad laser light is shone on the side facet of stacked waveguides separated with a cladding layer, where the waveguides act as a sensing layer and a reference layer that produce an interference pattern in a decaying (evanescent) electric field.

A full representation of DPI theory and instrumentation is shown in Figure 3.5 and Figure 3.6, respectively.

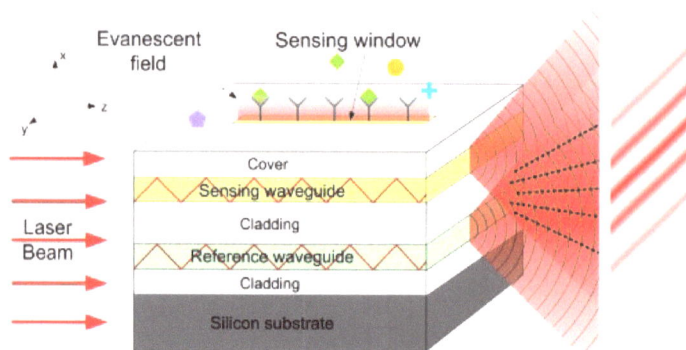

Figure 3.4: Basic representation of a slab waveguide interferometer. Reprinted from M. Wang, S. Uusitalo, C. Liedert, J. Hiltunen, L. Hakalahti, and R. Myllya, Polymeric dual-slab waveguide interferometer for biochemical sensing applications. *Appl. Optics,* **2012, 12, 1886. Copyright: The Optical Society (2012).**

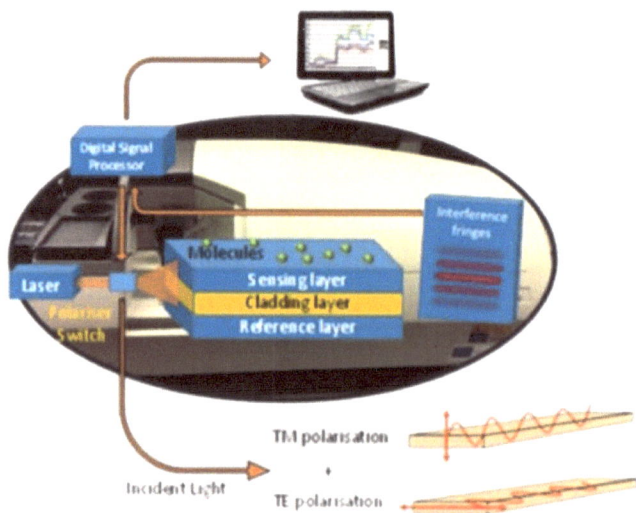

Figure 3.5: DPI sensing apparatus and fringe pattern collection from transverse-magnetic and transverse-electric polarizations of light. Adapted from J. Escorihuela, M. A. Gonzalez-Martinez, J. L. Lopez-Paz, R. Puchades, A. Maquieira, and D. Gimenez-Romero, Dual-polarization interferometry: a novel technique to light up the nanomolecular world. *Chem. Rev.,* **2015, 115, 265. Copyright: American Chemical Society (2015).**

The layer thickness and refractive index measurements are determined by measuring two phase changes in the system simultaneously because both transverse-electric and transverse-magnetic polarizations are allowed through the waveguides. Phase changes in each polarization of the light wave are lateral shifts of the wave peak from a given reference peak. The phase shifts are caused by changes in refractive index and layer thickness that result from molecular fluctuations in the sample. Switching between transverse-electric and transverse-magnetic polarizations happens very rapidly at 2 ms, where the switching mechanism is performed by a liquid crystal wave plate. This enables real-time measurements of the parameters to be obtained simultaneously.

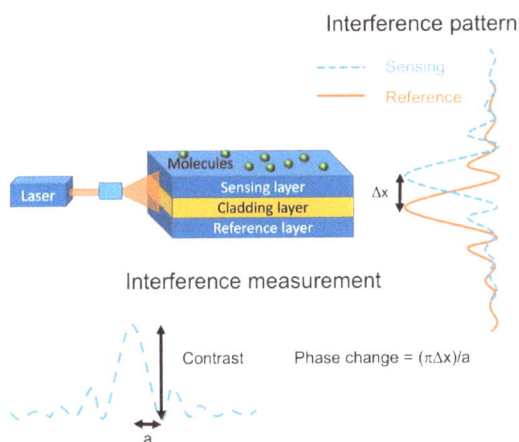

Figure 3.6: Fringe pattern detection of the waveguides and phase change determination between the sensing and reference interference patterns. Adapted from J. Escorihuela, M. A. Gonzalez-Martinez, J. L. Lopez-Paz, R. Puchades, A. Maquieira, and D. Gimenez-Romero, Dual-polarization interferometry: a novel technique to light up the nanomolecular world. *Chem. Rev.*, 2015, 115, 265. Copyright: American Chemical Society (2015).

Comparison of DPI with other techniques

Initial DPI evaluations

The first techniques rigorously compared to DPI were neutron reflection (NR) and X-ray diffraction. These studies demonstrated that DPI had a very high precision of 40 pm, which is comparable to NR and superior to X-ray diffraction. Additionally, DPI can provide real time information and conditions similar to an in-vivo environment, and the instrumental requirements are far simpler than those for NR. However, NR and X-ray diffraction are able to provide structural information that DPI cannot determine.

DPI comparison with orthogonal analytical techniques

Comparisons between DPI and alternative techniques have been performed since the initial evaluations, with techniques including surface plasmon resonance (SPR), atomic force microscopy (AFM), and quartz crystal microbalance with dissipation monitoring (QCM-D).

SPR is well-established for characterizing protein adsorption and has been used before DPI was developed. These techniques are very similar in that they both use an optical element based on an evanescent field, but they differ greatly in the method of calculating the mass of adsorbed protein. Rigorous testing showed that both tests give very accurate results, but their strengths differ. Because SPR uses spot-testing with an area of 0.26 mm^2 while DPI uses the average measurements over the length of the entire 15 mm chip, SPR is recommended for use in kinetic studies where diffusion in involved. However, DPI shows much greater accuracy than SPR when measuring refractive index and layer thickness.

Atomic Force Microscopy is a very different analytical technique than DPI because it is a type of microscopy used for high-resolution surface characterization. Hence, AFM and DPI are well-suited to be used in conjunction because AFM can provide accurate molecular structures and surface mapping while DPI provides layer thickness that AFM cannot determine.

QCM-D is a technique that can be used with DPI to provide complementary data. QCM-D differs from DPI by calculating both mass of the solvent and the mass of the adsorbed protein layer. These techniques can be used together to determine the amount of hydration in the adsorbed layer. QCM-D can also quantify the supramolecular conformation of the adlayer using energy dissipation calculations, while DPI can detect these conformational changes using birefringence, thus making these techniques orthogonal. One way that DPI is superior to QCM-D is that the latter techniques loses accuracy as the film becomes very thin, while DPI retains accuracy throughout the angstrom scale.

A tabulated representation of these techniques and their ability to measure structural detail, *in-vivo* conditions, and real time data is shown in Table 3.1.

Technique	Real time	Close to *in-vivo*[a]	Structural details
QCM-D	Yes	Yes	Medium
SPR	Yes	Yes	Low
X-ray	No	No	Very high
AFM	No	No	High
NR	No	Yes	(High
DPI	Yes	Yes	Medium

Table 3.1: Comparison of DPI with other analytical techniques. [a]Close to in-vivo means that the sensor can provide information that is similar to those experiences under in-vivo conditions. Data from J. Escorihuela, M. A. Gonzalez-Martinez, J. L. Lopez-Paz, R. Puchades, A. Maquieira, and D. Gimenez-Romero, Dual-polarization interferometry: a novel technique to light up the nanomolecular world. *Chem. Rev.*, 2015, 115, 265. Copyright: American Chemical Society (2015).

Applications of DPI

Protein Studies

DPI has been most heavily applied to protein studies. It has been used to elucidate membrane crystallization, protein orientation in a membrane, and conformational changes. It has also been used to study protein-protein interactions, protein-antibody interactions, and the stoichiometry of binding events.

Thin film studies

Since its establishment using protein interaction studies, DPI has seen its applications expanded to include thin film studies. DPI was compared to ellipsometry and QCM-D studies to indicate that it can be applied to heterogeneous thin films by applying revised analytical formulas to estimate the thickness, refractive index, and extinction coefficient of heterogeneous films that absorb light. A non-uniform density distribution model was developed and tested on polyethylenimine (Figure 3.7) deposited onto silica and compared to QCD-M measurements. Additionally, this revised model was able to calculate the mass of multiple species of molecules in composite films, even if the molecules absorbed different amounts of light. This information is valuable for providing surface composition.

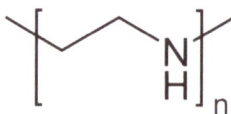

Figure 3.7: Structure of polyethylenimine used to form a thin film for DPI measurements.

A challenge of measuring layer thickness in thin films such as polyethylenimine is that DPI's evanescent field will create inaccurate measurements in inhomogeneous films as the film thickness increases. An error of approximately 5% was seen when layer thickness was increased to 90 nm. Data from this study determining the densities throughout the polyethylenimine film are shown in Figure 3.8.

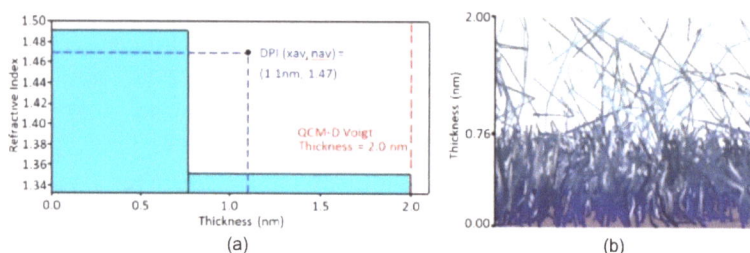

(a)   (b)

Figure 3.8: Density distribution of a polyethylenimine film using heterogeneous layer equations for DPI and QCM-D. Adapted from P. D. Coffey, M. J. Swann, T. A. Waigh, Q. Mua, and J. R. Lu, The structure and mass of heterogeneous thin films measured with dual polarization interferometry and ellipsometry. *RSC Adv.*, **2013, 3, 3316. Copyright: Royal Society of Chemistry (2013).**

Thin layer adsorption studies

Similar to thin film characterization studies, thin layers of adsorbed polymers have also been elucidated using DPI. It has been demonstrated that two different adsorption conformations of polyacrylamide form on resin, which provides useful information about adsorption behaviors of the polymer. This information is industrially important because polyacrylamide is widely used through the oil industry, and the adsorption of polyacrylamide onto resin is known to affect the oil/water interfacial stability.

Initial adsorption kinetics and conformations were also illuminated using DPI on bottlebrush polyelectrolytes. Bottlebrush polyelectrolytes are show in

Figure 3.9. It was shown that polyelectrolytes with high charge density initially adsorbed in layers that were parallel to the surface, but as polyelectrolytes were replaced with low charge density species, alignment changed to prefer perpendicular arrangement to the surface.

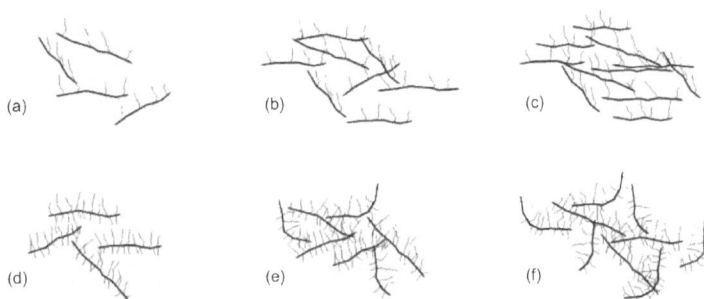

Figure 3.9: A representation of bottlebrush polyelectrolytes and how they adsorb to a layer differently over time as determined by DPI. Adapted from G. Bijelic, A. Shovsky, I. Varga, R. Makuska, and P. M. Claesson, Adsorption characteristics of brush polyelectrolytes on silicon oxynitride revealed by dual polarization interferometry. *J. Colloid Interface Sci.*, 2010, 348, 189. Copyright: Elesevier (2010).

Hg^{2+} biosensing studies

In 2009, it was shown by Wang et al. that DPI could be used for small molecule sensing. In their first study describing this use of DPI, they used single stranded DNA that was rich in thymine to complex Hg^{2+} ions. When DNA complexed with Hg^{2+}, the DNA transformed from a random coil structure to a hairpin structure. This change in structure could be detected by DPI at Hg^{2+} concentrations smaller than the threshold concentration allowed in drinking water, indicating the sensitivity of this label-free method for Hg^{2+} detection. High selectivity was indicated when the authors did not observe similar structural changes for Mg^{2+}, Ca^{2+}, Mn^{2+}, Fe^{3+}, Cd^{2+}, Co^{2+}, Ni^{2+}, Zn^{2+} or Pb^{2+} ions. A graphical description of this experiment is shown in Figure 3.10. It has been demonstrated that biosensing of small molecules and other metal cations can be achieved using other forms of functionalized DNA that specifically bind the desired analytes. Examples of molecules detected in this manner are shown in Figure 3.11.

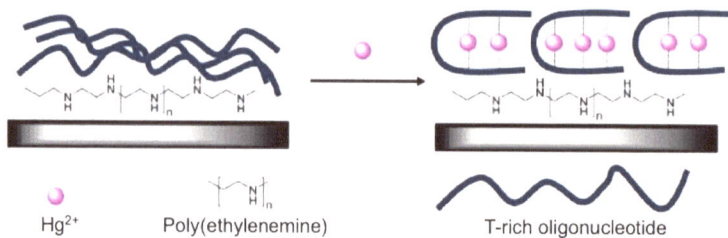

Hg²⁺   Poly(ethylenemine)   T-rich oligonucleotide

Figure 3.10: Selective Hg²⁺ detection using single strand DNA to complex the cation and measure the conformational changes in the DNA with DPI. Adapted from J. Escorihuela, M. A. Gonzalez-Martinez, J. L. Lopez-Paz, R. Puchades, A. Maquieira, and D. Gimenez-Romero, Dual-polarization interferometry: a novel technique to light up the nanomolecular world. *Chem. Rev.*, 2015, 115, 265. Copyright: American Chemical Society (2015).

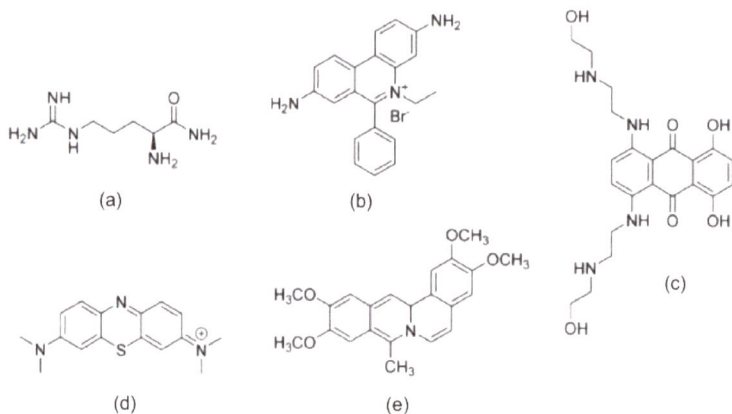

(a)   (b)   (c)   (d)   (e)

Figure 3.11: Small molecules detected using DPI measurements of functionalized DNA biosensors. Adapted from J. Escorihuela, M. A. Gonzalez-Martinez, J. L. Lopez-Paz, R. Puchades, A. Maquieira, and D. Gimenez-Romero, Dual-polarization interferometry: a novel technique to light up the nanomolecular world. *Chem. Rev.*, 2015, 115, 265. Copyright: American Chemical Society (2015).

Bibliography

G. Bijelic, A. Shovsky, I. Varga, R. Makuska, and P. M. Claesson, Adsorption characteristics of brush polyelectrolytes on silicon oxynitride revealed by dual polarization interferometry. *J. Colloid Interface Sci.*, 2010, **348**, 189.

P. D. Coffey, M. J. Swann, T. A. Waigh, Q. Mua, and J. R. Lu, The structure and mass of heterogeneous thin films measured with dual polarization interferometry and ellipsometry. *RSC Adv.*, 2013, **3**, 3316.

G. H. Cross, A. A. Reeves, S. Brand, J. F. Popplewell, L. L. Peel, M. J. Swann, and N. J. Freeman, A new quantitative optical biosensor for protein characterisation. *Biosens. Bioelectron.*, 2003, **19**, 383.

G. H. Cross, Y. Ren, and N. J. Freeman, Young's fringes from vertically integrated slab waveguides: Applications to humidity sensing. *J. Appl. Phys.*, 1999, **86**, 6483.

K. Li, M. Duan, H. Wang, J. Zhang, and B. Jing, Investigation on the adsorption behavior of polyacrylamide on resin by dual polarization interferometry. *RSC Adv.*, 2015, **5**, 17389.

J. Escorihuela, M. A. Gonzalez-Martinez, J. L. Lopez-Paz, R. Puchades, A. Maquieira, and D. Gimenez-Romero, Dual-polarization interferometry: a novel technique to light up the nanomolecular world. *Chem. Rev.*, 2015, **115**, 265.

N. J. Freeman and G. H. Cross, Planar waveguide chemical sensor, US Patent 6,335,793 (2002).

A. W. Sonesson, T.H. Callisen, H. Brismar, and U.M. Elofsson, A comparison between dual polarization interferometry (DPI) and surface plasmon resonance (SPR) for protein adsorption studies. *Colloids Surf. B*, 2007, **54**, 236.

M. J. Swann, L. L. Peel, S. Carrington, and N. J. Freeman, Dual-polarization interferometry: an analytical technique to measure changes in protein structure in real time, to determine the stoichiometry of binding events, and to differentiate between specific and nonspecific interactions. *Anal. Biochem.*, 2004, **329**, 190.

M. Wang, S. Uusitalo, C. Liedert, J. Hiltunen, L. Hakalahti, and R. Myllyla, Polymeric dual-slab waveguide interferometer for biochemical sensing applications. *Appl. Optics*, 2012, **12**, 1886.

Chapter 4: Scanning Electron Microscopy

Stacy Prukop and Andrew R. Barron

Introduction

The scanning electron microscope (SEM) is a very useful imaging technique that utilized a beam of electrons to acquire high magnification images of specimens. Very similar to the transmission electron microscope (TEM), the SEM maps the reflected electrons and allows imaging of thick (~mm) samples, whereas the TEM requires extremely thin specimens for imaging; however, the SEM has lower magnifications. Although both SEM and TEM use an electron beam, the image is formed very differently and users should be aware of when each microscope is advantageous.

Microscopy physics

Image formation

All microscopes serve to enlarge the size of an object and allow people to view smaller regions within the sample. Microscopes form optical images and although instruments like the SEM have extremely high magnifications, the physics of the image formation are very basic. The simplest magnification lens can be seen in Figure 4.1. The formula for magnification is,

$$M = f u - f = v - f f$$

where M is magnification, f is focal length, u is the distance between object and lens, and v is distance from lens to the image.

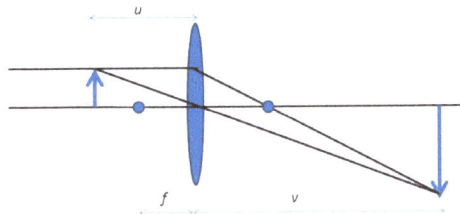

Figure 4.1: Basic microscope diagram illustrating inverted image and distances u, f, and v.

Multistage microscopes can amplify the magnification of the original object even more as shown in Figure 4.2. Where magnification is now calculated from,

$$M = \frac{(v_1 - f_1)(v_2 - f_2)}{f_1 f_2}$$

where, f_1, f_2 are focal distances with respect to the first and second lens and $v1$, $v2$ are the distances from the lens to the magnified image of first and second lens, respectively.

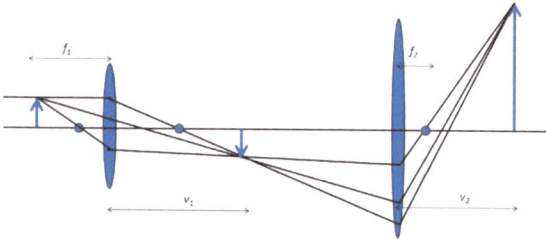

Figure 4.2: A schematic diagram of the optics used in a multistage microscope.

In reality, the objects we wish to magnify need to be illuminated. Whether or not the sample is thin enough to transmit light divides the microscope into two arenas. SEM is used for samples that do not transmit light, whereas the TEM (transmission electron microscope) requires transparent samples. Due to the many frequencies of light from the introduced source, a condenser system is added to control the brightness and narrow the range of viewing to reduce aberrations, which distort the magnified image.

Electron microscopes

Microscope images can be formed instantaneous (as in the optical microscope or TEM) or by rastering (scanning) a beam across the sample and forming the image point-by-point. The latter is how SEM images are formed. It is important to understand the basic principles behind SEM that define properties and limitations of the image.

Resolution

The resolution of a microscope is defined as the smallest distance between two features that can be uniquely identified (also called resolving power). There are many limits to the maximum resolution of the SEM and other microscopes, such as imperfect lenses and diffraction effects. Each single beam of light, once passed through a lens, forms a series of cones called an *airy ring* (see Figure 4.3). For a given wavelength of light, the central spot size is inversely proportional to the aperture size (i.e., large aperture yields small spot size) and high resolution demands a small spot size.

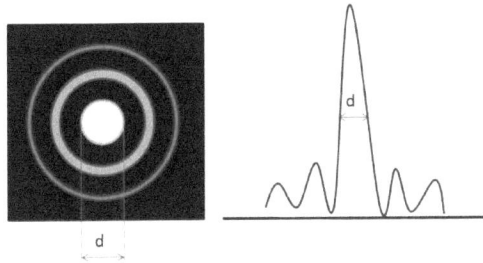

Figure 4.3: Airy ring illustrating center intensity (left) and intensity as a function of distance (right).

Aberrations distort the image and we try to minimize the effect as much as possible. Chromatic aberrations are caused by the multiple wavelengths present in *white* light. Spherical aberrations are formed by focusing inside and outside the ideal focal length and caused by the imperfections within the objective lenses. Astigmatism is because of further distortions in the lens. All aberrations decrease the overall resolution of the microscope.

Electrons

Electrons are charged particles and can interact with air molecules therefore the SEM and TEM instruments require extremely high vacuum to obtain images (10^{-7} atm). High vacuum ensures that very few air molecules are in the electron beam column. If the electron beam interacts with an air molecule, the air will become ionized and damage the beam filament, which is very costly to repair. The charge of the electron allows scanning and also inherently has a very small deflection angle off the source of the beam.

The electrons are generated with a thermionic filament. A tungsten (W) or LaB_6 filament (Figure 4.4) is chosen based on the needs of the user. LaB_6 is much more expensive and tungsten filaments meet the needs of the average user. The microscope can be operated as field emission (tungsten filament).

Figure 4.4: A lanthanum hexaboride (LaB_6) cathode.

Electron scattering

To accurately interpret electron microscopy images, the user must be familiar with how high energy electrons can interact with the sample and how these interactions affect the image. The probability that a particular electron will be scattered in a certain way is either described by the cross section, σ, or mean free path, λ, which is the average distance which an electron travels before being scattered.

Elastic scatter

Elastic scatter, or Rutherford scattering, is defined as a process which deflects an electron but does not decrease its energy. The wavelength of the scattered electron can be detected and is proportional to the atomic number. Elastically scattered electrons have significantly more energy that other types and provide mass contrast imaging. The mean free path, λ, is larger for smaller atoms meaning that the electron travels farther.

Inelastic scatter

Any process that causes the incoming electron to lose a detectable amount of energy is considered inelastic scattering. The two most common types of inelastic scatter are phonon scattering and plasmon scattering. Phonon scattering occurs when a primary electron loses energy by exciting a phonon, atomic vibrations in a solid, and heats the sample a small amount. A Plasmon is an oscillation within the bulk electrons in the conduction band for metals. Plasmon scattering occurs when an electron interacts with the sample and produces plasmons, which typically have 5 - 30 eV energy loss and small λ.

Secondary effects

A secondary effect is a term describing any event which may be detected *outside* the specimen and is essentially how images are formed. To form an image, the electron must interact with the sample in one of the aforementioned ways and *escape* from the sample and be detected. *Secondary electrons* (SE) are the most common electrons used for imaging due to high abundance and are defined, rather arbitrarily, as electrons with less than 50 eV energy after exiting the sample. *Backscattered electrons* (BSE) leave the sample quickly and retain a high amount of energy; however, there is a much lower yield of BSE. Backscattered electrons are used in many different imaging modes. Refer to Figure 4.5 for a diagram of interaction depths corresponding to various electron interactions.

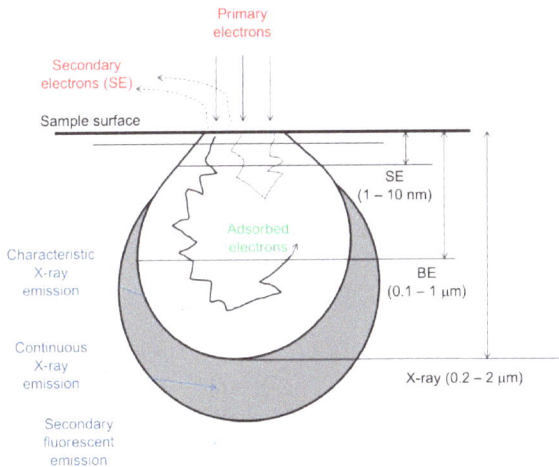

Figure 4.5 Diagram illustrating the depths at which various sample interactions occur.

SEM Construction

The SEM is made of several main components: electron gun, condenser lens, scan coils, detectors, specimen, and lenses (see Figure 4.6). Today, portable SEMs are available, but the typical size is about 6 feet tall and contains the microscope column and the control console.

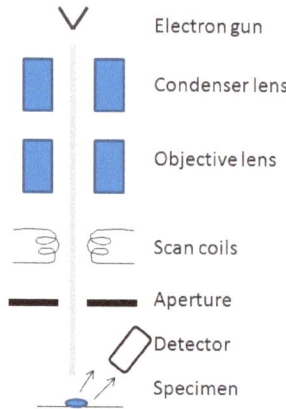

Figure 4.6: Schematic drawing of the SEM illustrating placement of electron generation, collimation process, sample interaction and electron detection.

A special feature of the SEM and TEM is known as *depth of focus, dv/du* the range of positions (depths) at which the image can be viewed with good focus,

$$dv/du = -v^2/u^2 = -M^2$$

This allows the user to see more than a singular plane of a specified height in focus and essentially allows a range of three-dimensional imaging.

Electron detectors (image formation)

The secondary electron detector (SED) is the main source of SEM images since a large majority of the electrons emitted from the sample are less than 50 eV. These electrons form textural images but cannot determine composition. The SEM may also be equipped with a backscatter electron detector (BSED) which collects the higher energy BSE's. Backscattered electrons are very sensitive to atomic number and can determine qualitative information about nuclei present (i.e., how much Fe is in the sample). Topographic images are taken by tilting the specimen 20 - 40° toward the detector. With the sample tilted, electrons are more likely to scatter off the top of the sample rather than interact within it, thus yielding information about the surface.

Sample preparation

The most effective SEM sample will be at least as thick as the interaction volume; depending on the image technique you are using (typically at least 2 μm). For the best contrast, the sample must be conductive, or the sample can be sputter-coated with a metal (such as Au, Pt, W, and Ti). Metals and other materials that are naturally conductive do not need to be coated and need very little sample preparation.

SEM of polymers

As previously discussed, to view features that are smaller than the wavelength of light, an electron microscope must be used. The electron beam requires extremely high vacuum to protect the filament and electrons must be able to adequately interact with the sample. Polymers are typically long chains of repeating units composed primarily of "lighter" (low atomic number) elements such as carbon, hydrogen, nitrogen, and oxygen. These lighter elements have fewer interactions with the electron beam which yields poor contrast, so often times a stain or coating is required to view polymer samples. SEM imaging requires a conductive surface, so a large majority of polymer samples are sputter coated with metals, such as gold.

The decision to view a polymer sample with an SEM (versus a TEM for example) should be evaluated based on the feature size you expect the sample to have. Generally, if you expect the polymer sample to have features, or even individual molecules, over 100 nm in size you can safely choose SEM to view your sample. For much smaller features, the TEM may yield better results, but requires much different sample preparation than will be described here.

Polymer sample preparation techniques

Sputter coating

A sputter coater may be purchased that deposits single layers of gold, gold-palladium, tungsten, chromium, platinum, titanium, or other metals in a very controlled thickness pattern. It is possible, and desirable, to coat only a few nm's of metal onto the sample surface.

Spin coating

Many polymer films are depositing via a spin coater which spins a substrate (often ITO glass) and drops of polymer liquid are dispersed an even thickness on top of the substrate.

Staining

Another option for polymer sample preparation is staining the sample. Stains act in different ways, but typical stains for polymers are osmium tetroxide (OsO_4, Figure 4.7), ruthenium tetroxide (RuO_4) phosphotungstic acid ($H_3PW_{12}O_{40}$, Figure 4.78 hydrazine (N_2H_4), and silver sulfide (Ag_2S).

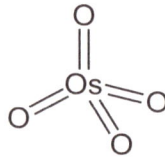

Figure 4.7: Structure of osmium tetroxide (OsO_4).

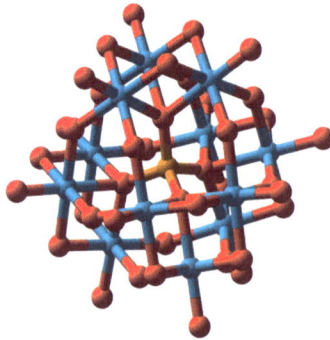

Figure 4.8: Structure of phosphotungstic acid ($H_3PW_{12}O_{40}$), where blue = tungsten, red = oxygen, and orange = phosphorus

Examples

Comb-block copolymer (microstructure of cast film)

- Cast polymer film (see Figure 4.9).
- To view interior structure, the film was cut with a microtome or razor blade after the film was frozen in liquid N_2 and fractured.

- Stained with RuO_4 vapor (after cutting).
- Structure measurements were averaged over a minimum of 25 measurements.

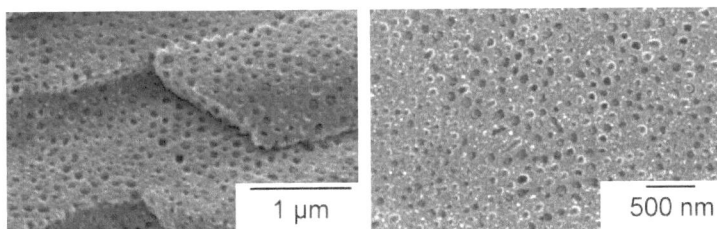

Figure 4.9: SEM micrograph of comb block copolymer showing spherical morphology and long range order. Adapted from M. B. Runge and N. B. Bowden, Synthesis of high molecular weight comb block copolymers and their assembly into ordered morphologies in the solid state. *J. Am. Chem. Soc.*, **2007, 129, 10551. Copyright: American Chemical Society (2007).**

Polystyrene-polylactide bottlebrush copolymers (lamellar spacing)

- Pressed polymer samples into disks and annealed for 16 h at 170 °C.
- To determine ordered morphologies, the disk was fractured (Figure 4.10).
- Used SEM to verify lamellar spacing from USAXS.

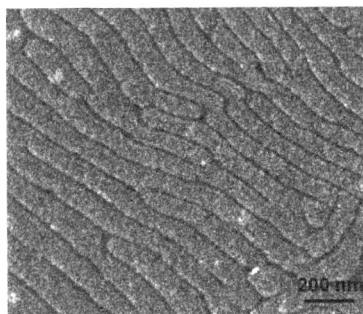

Figure 4.10: SEM image of a fractured piece of polymer SL-1. Adapted from J. Rzayev, Synthesis of polystyrene−polylactide bottlebrush block copolymers and their melt self-assembly into large domain nanostructures. *Macromolecules*, **2009, 42, 2135. Copyright: American Chemical Society (2009).**

SWCNTs in ultra-high molecular weight polyethylene

- Dispersed single walled carbon nanotubes (SWCNTs) in interactive polyethylene polymer.
- Samples were sputter-coated in gold to enhance contrast.
- The films were solution-crystallized, and the cross-section was imaged.
- Environmental SEM (ESEM) was used to show morphologies of composite materials.
- WD = 7 mm.
- Study was conducted to image sample before and after drawing of film.
- Images confirmed the uniform distribution of SWNT in PE (Figure 4.11).
- M_W = 10,000 Dalton.
- Study performed to compare transparency before and after UV irradiation.

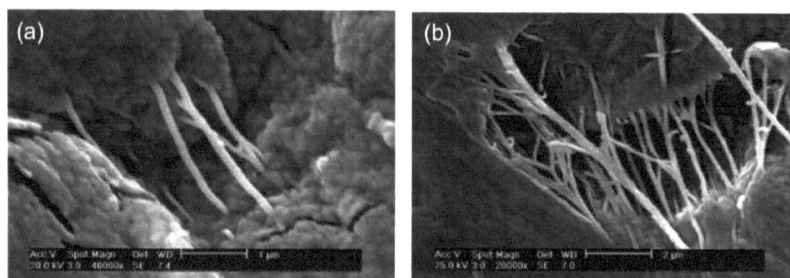

Figure 4.11: SEM images of crystallized SWNT-UHMWPE films before (a) and after (b) drawing at 120 °C. Adapted from Q. Zhang, D. R. Lippits, and S. Rastogi, Dispersion and rheological aspects of SWNTs in ultrahigh molecular weight polyethylene. *Macromolecules*, 2006, 39, 658. Copyright: American Chemical Society (2006).

Nanostructures in conjugated polymers (nanoporous films)

- Polymer and NP were processed into thin films and heated to crosslink.
- SEM was used to characterize morphology and crystalline structure (Figure 4.11).
- SEM was used to determine porosity and pore size.

- Magnified orders of 200 nm - 1 μm.
- WD = 8 mm.
- M_W = 23,000 Daltons
- Sample prep: spin coating a solution of poly-(thiophene ester) with copper NPs suspended on to ITO coated glass slides. Ziess, Supra 35

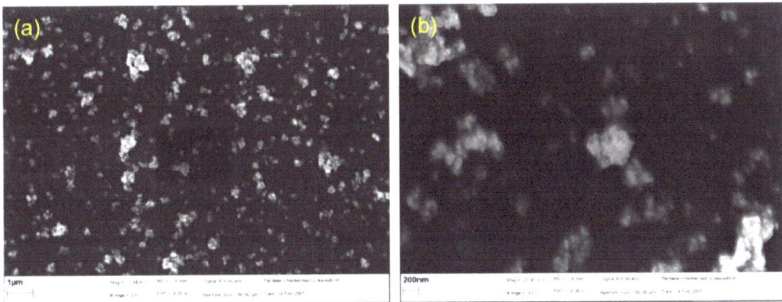

Figure 4.11: SEM images of thermocleaved film loaded with nanoparticles with scale bar 1 μm. Adapted from J. W. Andreasen, M. Jorgensen, and F. C. Krebs, A route to stable nanostructures in conjugated polymers. *Macromolecules***, 2007, 40, 7758. Copyright: American Chemical Society (2007).**

Cryo-SEM colloid polystyrene latex particles (fracture patterns)

- Used cryogenic SEM (cryo-SEM) to visualize the microstructure of particles (Figure 4.12).
- Particles were immobilized by fast-freezing in liquid N_2 at –196 °C.
- Sample is fractured (-196 °C) to expose cross section.
- 3 nm sputter coated with platinum.
- Shapes of the nanoparticles after fracture were evaluated as a function of crosslink density.

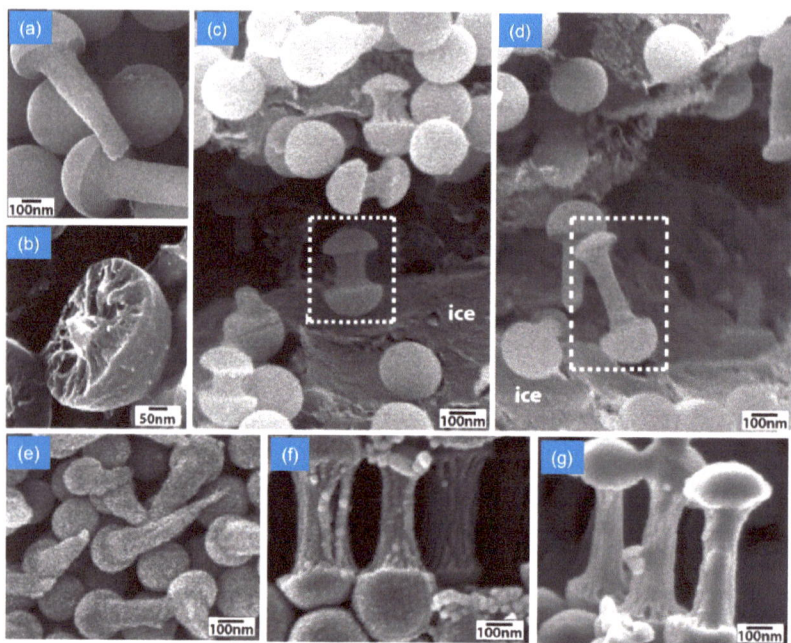

Figure 4.12: Cryo-SEM images of plastically drawn polystyrene and latex particles. Adapted from H. Ge, C. L. Zhao, S. Porzio, L. Zhuo, H. T. Davis, and L. E. Scriven, Fracture behavior of colloidal polymer particles in fast-frozen suspensions viewed by cryo-SEM. *Macromolecules*, **2006, 39, 5531. Copyright: American Chemical Society (2006).**

Bibliography

J. W. Andreasen, M. Jorgensen, and F. C. Krebs, A route to stable nanostructures in conjugated polymers. *Macromolecules*, 2007, **40**, 7758.

H. Ge, C. L. Zhao, S. Porzio, L. Zhuo, H. T. Davis, and L. E. Scriven, Fracture behavior of colloidal polymer particles in fast-frozen suspensions viewed by cryo-SEM. *Macromolecules*, 2006, **39**, 5531.

P. J. Goodhew, J. Humphreys, and R. Beanland, *Electron Microscopy and Analysis*, Taylor & Francis Inc., New York (2001).

R. A. Horch, N. Shahid, A. S. Mistry, M. D. Timmer, A. G. Mikos, and A. R. Barron, Reinforcement of poly(propylene fumarate)-based networks with surface modified alumoxane nanoparticles for bone tissue engineering. *Biomacromolecules*, 2004, **5**, 1990.

Y. Koide and A. R. Barron, Polyketone polymers prepared using a palladium/alumoxane catalyst system. *Macromolecules,* 1996, **29**, 1110

M. B. Runge and N. B. Bowden, Synthesis of high molecular weight comb block copolymers and their assembly into ordered morphologies in the solid state. *J. Am. Chem. Soc.*, 2007, **129**, 10551.

J. Rzayev, Synthesis of polystyrene–polylactide bottlebrush block copolymers and their melt self-assembly into large domain nanostructures. *Macromolecules*, 2009, **42**, 2135.

N. Shahid, R. Villate, and A. R. Barron, Chemically functionalized alumina nanoparticle effect on carbon fiber/epoxy composites. *Composite Sci. Tech.*, 2005, **65**, 2250

C. T. Vogelson, Y. Koide, R. Cook, S. G. Bott, L. B. Alemany, and A. R. Barron, Inorganic-organic hybrid and composite resin materials using carboxylate-alumoxanes as functionalized cross-linking agents. *Chem. Mater.*, 2000, **12**, 795

Q. Zhang, D. R. Lippits, and S. Rastogi, Dispersion and rheological aspects of SWNTs in ultrahigh molecular weight polyethylene. *Macromolecules*, 2006, **39**, 658.

Chapter 5: Transmission Electron Microscopy

Zhengzong Sun, Dayne Swearer, Yen-Tien Lu, Zhe Wang,
Pavan M. V. Raja and Andrew R. Barron

Introduction

Transmission electron microscopy (TEM) is a form of microscopy which in which a beam of electrons transmits through an extremely thin specimen, and then interacts with the specimen when passing through it. The formation of images in a TEM can be explained by an optical electron beam diagram in Figure 5.1. TEMs provide images with significantly higher resolution than visible-light microscopes (VLMs) do because of the smaller de Broglie wavelength of electrons. These electrons allow for the examination of finer details, which are several thousand times higher than the highest resolution in a VLM. Nevertheless, the magnification provide in a TEM image is in contrast to the absorption of the electrons in the material, which is primarily due to the thickness or composition of the material.

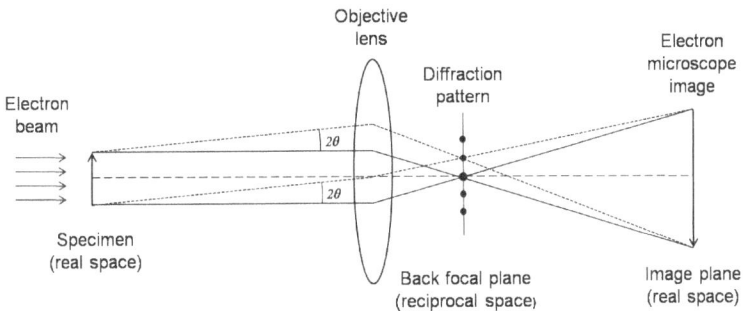

Figure 5.1: The optical electron beam diagram of TEM.

When a crystal lattice spacing (d) is investigated with electrons with wavelength λ, diffracted waves will be formed at specific angles 2θ, satisfying the Bragg condition,

$$2d\sin\theta = \lambda$$

The regular arrangement of the diffraction spots, the so-called diffraction pattern (DP), can be observed. While the transmitted and the diffracted beams interfere on the image plane, a magnified image (electron microscope image) appears. The plane where the DP forms is called the *reciprocal space*, which the image plane is called the *real space*. A Fourier transform can mathematically transform the real space to reciprocal space.

By adjusting the lenses (changing their focal lengths), both electron microscope images and DP can be observed. Thus, both observation modes can be successfully combined in the analysis of the microstructures of materials. For instance, during investigation of DPs, an electron microscope image is observed. Then, by inserting an aperture (selected area aperture), adjusting the lenses, and focusing on a specific area that we are interested in, we will get a DP of the area. This kind of observation mode is called a *selected area diffraction*. In order to investigate an electron microscope image, we first observe the DP. Then by passing the transmitted beam or one of the diffracted beams through a selected aperture and changing to the imaging mode, we can get the image with enhanced contrast, and precipitates and lattice defects can easily be identified.

Describing the resolution of a TEM in terms of the classic Rayleigh criterion for VLMs, which states that the smallest distance that can be investigated, δ, is given approximately by

$$\delta = 0.61\lambda/(\mu\sin\beta)$$

where λ is the wavelength of the electrons, μ is the refractive index of the viewing medium, and β is the semi-angle of collection of the magnifying lens.

According to de Broglie's ideas of the wave-particle duality, the particle momentum p is related to its wavelength λ through Planck's constant h,

$$\lambda = h/p$$

Momentum is given to the electron by accelerating it through a potential drop, V, giving it a kinetic energy, eV. This potential energy is equal to the kinetic energy of the electron,

$$eV = (m_0 v^2)/2$$

Based upon the foregoing, we can equate the momentum (p) to the electron mass (m_o), multiplied by the velocity (v) and substituting for v,

$$p = m_0v = (2m_0eV)^{1/2}$$

These equations define the relationship between the electron wavelength, λ, and the accelerating voltage of the electron microscope (V), Eq. However, we have to consider about the relative effects when the energy of electron more than 100 keV. So, in order to be exact we must modify,

$$\lambda = \frac{h}{(2m_oeV)^{\frac{1}{2}}}$$

to give,

$$\lambda = \frac{h}{\left[2m_oeV\left(1 + \frac{eV}{2m_oe^2}\right)\right]^{\frac{1}{2}}}$$

If a higher resolution is desired a decrease in the electron wavelength is accomplished by increasing the accelerating voltage of the electron microscope. In other words, the higher accelerating rating used, the better resolution obtained.

Why the specimen should be thin

The scattering of the electron beam through the material under study can form different angular distribution (Figure 5.2) and it can be either forward scattering or back scattering. If an electron is scattered < 90°, then it is forward scattered, otherwise, it is backscattered. If the specimen is thicker, fewer electrons are forward scattered and more are backscattered. Incoherent, backscattered electrons are the only remnants of the incident beam for bulk, non-transparent specimens. The reason that electrons can be scattered through different angles is related to the fact that an electron can be scattered more than once. Generally, the more times of scattering happen, the greater the angle of scattering.

(a)   (b)

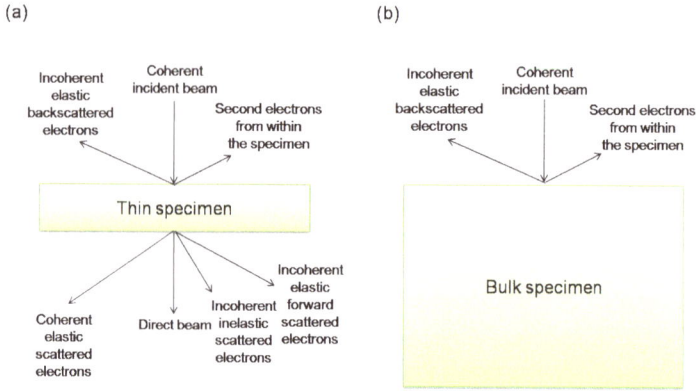

Figure 5.2: Two different kinds of electron scattering form (a) a thin specimen and (b) a bulk specimen.

All scattering in the TEM specimen is often approximated as a single scattering event since it is the simplest process. If the specimen is very thin, this assumption will be reasonable enough. If the electron is scattered more than once, it is called 'plural scattering.' It is generally safe to assume single scattering occurs, unless the specimen is particularly thick. When the times of scattering increase, it is difficult to predict what will happen to the electron and to interpret the images and DPs. So, the principle is 'thinner is better', i.e., if we make thin enough specimens so that the single-scattering assumption is plausible, and the TEM research will be much easier.

In fact, forward scattering includes the direct beam, most elastic scattering, refraction, diffraction, particularly Bragg diffraction, and inelastic scattering. Because of forward scattering through the thin specimen, a DP or an image would be showed on the viewing screen, and an X-ray spectrum or an electron energy-loss spectrum can be detected outside the TEM column. However, backscattering still cannot be ignored, it is an important imagine mode in the SEM.

Instrument

Although TEM can easily get to atomic resolution, the first TEM invented by Ernst Ruska (Figure 5.3) in April 1932 could hardly compete with optical microscope, with only 3.6×4.8 = 14.4 magnification. The primary problem was the electron irradiation damage to sample in poor vacuum system. After World War II, Ruska resumed his work in developing high resolution TEM.

Finally, this work brought him the Nobel Prize in physics 1986. Since then, the general structure of TEM hasn't changed too much as shown in Figure 5.4. The basic components in TEM are: electron gun, condenser system, objective lens (most important lens in TEM, which determines the final resolution), diffraction lens, projective lenses (all lens are inside the equipment column, between apertures), image recording system (used to be negative films, now is CCD cameras) and vacuum system.

Figure 5.3: German physicist Ernst August Friedrich Ruska (1906 - 1988) who won the Nobel Prize in Physics in 1986 for his work in electron optics, including the design of the first electron microscope.

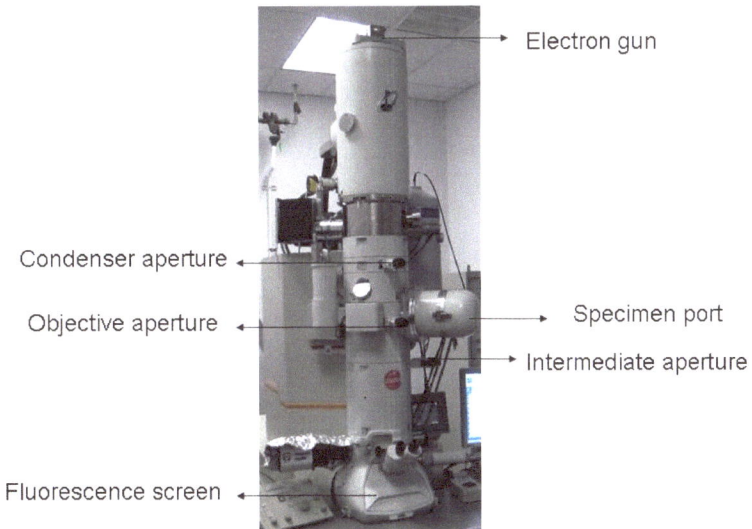

Figure 5.4: The basic components in a TEM instrument.

Limitations of TEM

Interpreting transmission images

One significant problem that might encounter when TEM images are analyzed is that the TEM present us with 2D images of a 3D specimen, viewed in transmission. This problem can be illustrated by showing a picture of two rhinos, side by side such that the head of one appears attached to the rear of the other (Figure 5.5).

Figure 5.5: In projection, this photograph of two rhinos appears as one two-headed beast, because sometimes people have difficulty to translate a 2D image to a 3D image. Adapted from D. B. Williams and C. B. Carter, *Transmission Electron Microscopy: A Textbook for Material Science,* **2nd edn., Springer, New York (2009).**

One aspect of this particular drawback is that a single TEM images has no depth sensitivity. There often is information about the top and bottom surfaces of the specimen, but this is not immediately apparent. There has been progress in overcoming this limitation, by the development of electron tomography, which uses a sequence of images taken at different angles. In addition, there has been improvement in specimen-holder design to permit full 360° rotation and, in combination with easy data storage and manipulation; nanotechnologists have begun to use this technique to look at complex 3D inorganic structures such as porous materials containing catalyst particles.

Electron beam damage

A detrimental effect of ionizing radiation is that it can damage the specimen, particularly polymers (and most organics) or certain minerals and ceramics. Some aspects of beam damage made worse at higher voltages. Figure 5.6 shows an area of a specimen damaged by high-energy electrons. However, the combination of more intense electron sources with more sensitive electron detectors, and the use computer enhancement of noisy images, can be used to minimize the total energy received by the sample.

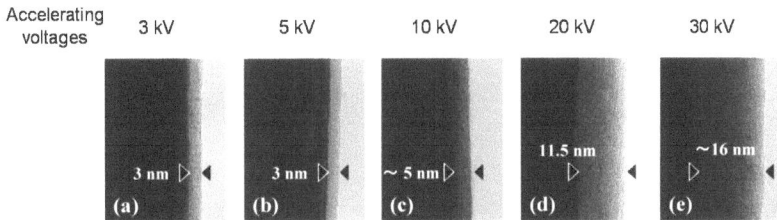

Figure 5.6: High-resolution TEM images at the slit edge of the GaAs samples prepared by slit focused ion beam. GaAs samples prepared at (a) 3 kV, (b) 5 kV, (c) 10 kV, (d) 20 kV, and (e) 30 kV. The thickness of the amorphous layer produced by focused ion beam is shown in each image. Adapted from Y. Yabuuchi, S. Tametou, T. Okano, S. Inazato, S. Sadayamn, and Y. Tamamoto, A study of the damage on FIB-prepared TEM samples of $Al_xGa_{1-x}As$. *J. Electron Microsc.*, 2004, 53, 5. Copyright: Oxford Academic (2004).

Sample preparation

The specimens under study have to be thin if any information is to be obtained using transmitted electrons in the TEM. For a sample to be transparent to electrons, the sample must be thin enough to transmit sufficient electrons such that enough intensity falls on the screen to give an image. This is a function of the electron energy and the average atomic number of the elements in the sample. Typically for 100 keV electrons, a specimen of aluminum alloy up to ~ 1 μm would be thin, while steel would be thin up to about several hundred nanometers. However, thinner is better and specimens < 100 nm should be used wherever possible.

The method to prepare the specimens for TEM depends on what information is required. In order to observe TEM images with high resolution, it is necessary to prepare thin films without introducing contamination or defects. For this purpose, it is important to select an appropriate specimen preparation method for each material, and to find an optimum condition for each method.

Crushing

A specimen can be crushed with an agate mortar and pestle. The flakes obtained are suspended in an organic solvent (e.g., acetone), and dispersed with a sonic bath or simply by stirring with a glass stick. Finally, the solvent containing the specimen flakes is dropped onto a grid. This method is limited to materials which tend to cleave (e.g., mica).

Electropolishing

Slicing a bulk specimen into wafer plates of about 0.3 mm thickness by a fine cutter or a multi-wire saw. The wafer is further thinned mechanically down to about 0.1 mm in thickness. Electropolishing is performed in a specific electrolyte by supplying a direct current with the positive pole at the thin plate and the negative pole at a stainless steel plate. In order to avoid preferential polishing at the edge of the specimen, all the edges are cover with insulating paint. This is called the window method. The electropolishing is finished when there is a small hole in the plate with very thin regions around it (Figure 5.7). This method is mainly used to prepare thin films of metals and alloys.

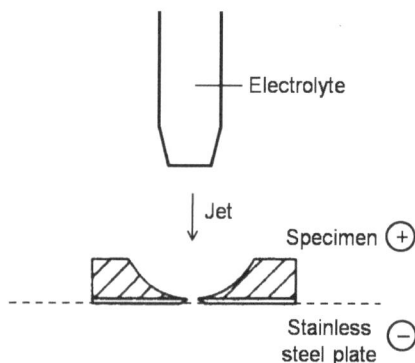

Figure 5.7: Principle of jet electropolishing method. The specimen and the stainless steel plate is electronic positive and negative, respectively.

Chemical polishing

Thinning is performed chemically, i.e., by dipping the specimen in a specific solution. As for electropolishing, a thin plate of 0.1 - 0.2 mm in thickness should be prepared in advance. If a small dimple is made in the center of the plate with a dimple grinder, a hole can be made by etching around the center while keeping the edge of the specimen relatively thick. This method is frequently used for thinning semiconductors such as silicon. As with electropolishing, if the specimen is not washed properly after chemical etching, contamination such as an oxide layer forms on the surface.

Ultramicrotomy

Specimens of thin films or powders are usually fixed in an acrylic or epoxy resin and trimmed with a glass knife before being sliced with a diamond knife. This process is necessary so that the specimens in the resin can be sliced easily by a diamond knife. Acrylic resins are easily sliced and can be removed with chloroform after slicing. When using an acrylic resin, a gelatin capsule is used as a vessel. Epoxy resin takes less time to solidify than acrylic resins, and they remain strong under electron irradiation. This method has been used for preparing thin sections of biological specimens and sometimes for thin films of inorganic materials which are not too hard to cut.

Ion milling

A thin plate (less than 0.1 mm) is prepared from a bulk specimen by using a diamond cutter and by mechanical thinning. Then, a disk 3 mm in diameter is made from the plate using a diamond knife or an ultrasonic cutter, and a dimple is formed in the center of the surface with a dimple grinder. If it is possible to thin the disk directly to 0.03 mm in thickness by mechanical thinning without using a dimple grinder, the disk should be strengthened by covering the edge with a metal ring. Ar ions are usually used for the sputtering, and the incidence angle against the disk specimen and the accelerating voltage are set as 10 - 20° and a few kilovolts, respectively. This method is widely used to obtain thin regions of ceramics and semiconductors in particular, and also for cross section of various multilayer films.

Focused ion beam (FIB)

This method was originally developed for the purpose of fixing semiconductor devices. In principle, ion beams are sharply focused on a small area, and the specimen in thinned very rapidly by sputtering. Usually Ga ions are used, with an accelerating voltage of about 30 kV and a current of about 10 A/cm^2. The probe size is several tens of nanometers. This method is useful for specimens containing a boundary between different materials, where it may be difficult to homogeneously thin the boundary region by other methods such as ion milling.

Vacuum evaporation

The specimen to be studied is set in a tungsten-coil or basket. Resistance heating is applied by an electric current passing through the coil or basket, and the specimen is melted, then evaporated (or sublimed), and finally deposited onto a substrate. The deposition process is usually carried under a pressure of 10$^-$^{3}-10^{-4} Pa, but in order to avoid surface contamination, a very high vacuum is necessary. A collodion film or cleaved rock salt is used as a substrate. Rock salt is especially useful in forming single crystals with a special orientation relationship between each crystal and the substrate. Salt is easily dissolved in water, and then the deposited films can be fixed on a grid. Recently, as an alternative to resistance heating, electron beam heating or an ion beam sputtering method has been used to prepare thin films of various alloys. This method is used for preparing homogeneous thin films of metals and alloys and is also used for coating a specimen with the metal of alloy.

The characteristics of the grid

The types of TEM specimens that are prepared depend on what information is needed. For example, a self-supporting specimen is one where the whole specimen consists of one material (which may be a composite). Other specimens are supported on a grid or on a Cu washer with a single slot. Some grids are shown in Figure 5.8. Usually the specimen or grid will be 3 mm in diameter.

TEM specimen stage designs include airlocks to allow for insertion of the specimen holder into the vacuum with minimal increase in pressure in other areas of the microscope. The specimen holders are adapted to hold a standard size of grid upon which the sample is placed or a standard size of self-supporting specimen. Standard TEM grid sizes is a 3.05 mm diameter ring, with a thickness and mesh size ranging from a few to 100 μm. The sample is placed onto the inner meshed area having diameter of approximately 2.5 mm. The grid materials usually are copper, molybdenum, gold or platinum. This grid is placed into the sample holder which is paired with the specimen stage. A wide variety of designs of stages and holders exist, depending upon the type of experiment being performed. In addition to 3.05 mm grids, 2.3 mm grids are sometimes, if rarely, used. These grids were particularly used in the mineral sciences where a large degree of tilt can be required and where specimen material may be extremely rare. Electron transparent specimens have a thickness around 100 nm, but this value depends on the accelerating voltage.

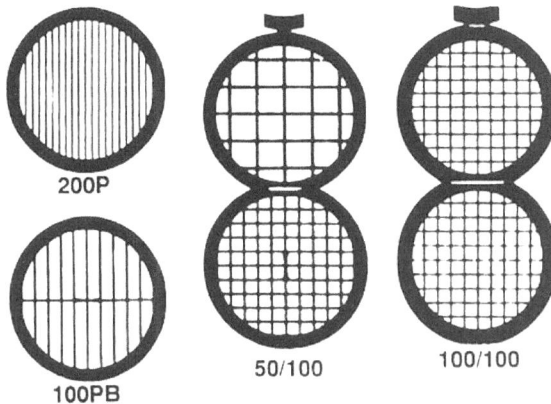

Figure 5.8: TEM sample support mesh grids. A diameter of a grid is usually 3.05 mm, however, some grids with diameters of 2.30 mm are also be used for earlier microscopes. Adapted from D. B. Williams and C. B. Carter, *Transmission Electron Microscopy: A Textbook for Material Science*, 2nd edn., Springer, New York (2009).

Once inserted into a TEM, the sample is manipulated to allow study of the region of interest. To accommodate this, the TEM stage includes mechanisms for the translation of the sample in the xy plane of the sample, for z height adjustment of the sample holder, and usually at least one rotation degree of freedom. Most TEMs provide the ability for two orthogonal rotation angles of movement with specialized holder designs called double-tilt sample holders.

A TEM stage is required to have the ability to hold a specimen and be manipulated to bring the region of interest into the path of the electron beam. As the TEM can operate over a wide range of magnifications, the stage must simultaneously be highly resistant to mechanical drift as low as a few nm/minute while being able to move several µm/minute, with repositioning accuracy on the order of nanometers.

Transmission electron microscopy image for multilayer-nanomaterials

Although, TEMs can only provide 2D analysis for a 3D specimen; magnifications of 300,000 times can be routinely obtained for many materials making it an ideal method for the study of nanomaterials. Besides from the TEM images, darker areas of the image show that the sample is thicker or denser in these areas, so we can observe the different components and structures of the specimen by the difference of color. For investigating multilayer-nanomaterials, a TEM is usually the first choice, because not only does it provide a high-resolution image for nanomaterials but also it can distinguish each layer within a nanostructured material.

Observations of multilayer-nanomaterials

TEM was been used to analyze the depth-graded W/Si multilayer films. Multilayer films were grown on polished, 100 mm thick Si wafers by magnetron sputtering in argon gas. The individual tungsten and silicon layer thicknesses in periodic and depth-graded multilayers are adjusted by varying the computer-controlled rotational velocity of the substrate platen. The deposition times required to produce specific layer thicknesses were determined from detailed rate calibrations. Samples for TEM were prepared by focused ion beam milling at liquid N_2 temperature to prevent any beam heating which might result in re-crystallization and/or re-growth of any amorphous or fine-grained polycrystalline layers in the film.

TEM measurements were made using a JEOL-4000 high-resolution transmission electron microscope operating at 400 keV; this instrument has a point-to-point resolution of 0.16 nm. Large area cross-sectional images of a depth-graded multilayer film obtained under medium magnification (~100 kX) were acquired at high resolution. A cross-sectional TEM image showed 150 layers W/Si film with the thickness of layers in the range of 3.33 ~ 29.6 nm (Figure 5.9 shows a part of layers). The dark layers are tungsten and the light layers are silicon and they are separated by the thin amorphous W–Si interlayers (gray bands). By the high resolution of the TEM and the nature characteristics of the material, each layer can be distinguished clearly with their different darkness.

Not all kinds of multilayer nanomaterials can be observed clearly under TEM. A material consists of pc-Si:H multilayers were prepared by a photo-assisted chemical vapor deposition (photo-CVD) using a low-pressure mercury lamp as an UV light source to dissociate the gases. The pc-Si:H multilayer included low H_2-diluted a-Si:H sublayers (SL's) and highly H_2-diluted a-Si:H sublayers (SH's). Control of the CVD gas flow ($H_2|SiH_4$) under continuous UV irradiation resulted in the deposition of multilayer films layer by layer.

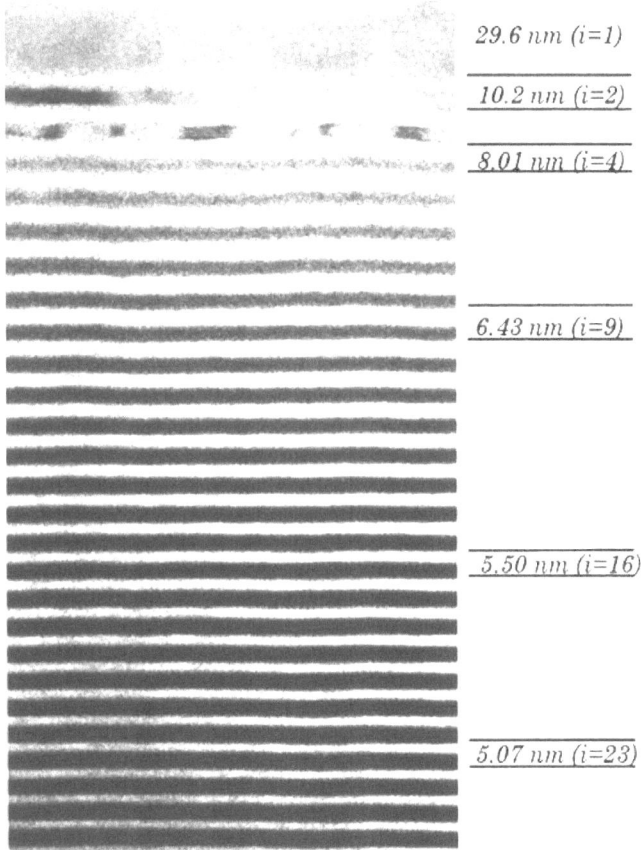

29.6 nm (i=1)

10.2 nm (i=2)

8.01 nm (i=4)

6.43 nm (i=9)

5.50 nm (i=16)

5.07 nm (i=23)

Figure 5.9: Cross-sectional transmission electron micrograph of the top portion of a depth-graded W/Si multilayer structure. Selected bilayer indices and thicknesses are indicated. The tungsten (dark bands) and silicon (light bands) layers are separated by thin amorphous W–Si interlayers (gray bands). The topmost silicon layer is not completely visible in this image. Adapted from D. L. Windt, F. E. Christensen, W. W. Craig, C. Hailey, F. A. Harrison, M. Jimenez-Garate, R. Kalyanaraman, and P. H. Mao, Growth, structure, and performance of depth-graded W/Si multilayers for hard X-ray optics. *J. Appl. Phys.*, 2000, 88, 460. Copyright: American Institute of Physics (2000).

For a TEM measurement, a 20 nm thick undiluted a-Si:H film on a c-Si wafer before the deposition of multilayer to prevent from any epitaxial growth. Figure 5.10 shows a cross-sectional TEM image of a six-cycled pc-Si:H multilayer specimen. The white dotted lines are used to emphasize the horizontal stripes, which have periodicity in the TEM image. As can be seen, there are no significant boundaries between SL and SH could be observed because all sublayers are prepared in H$_2$ gas. In order to get the more accurate thickness of each sublayer, other measurements might be necessary.

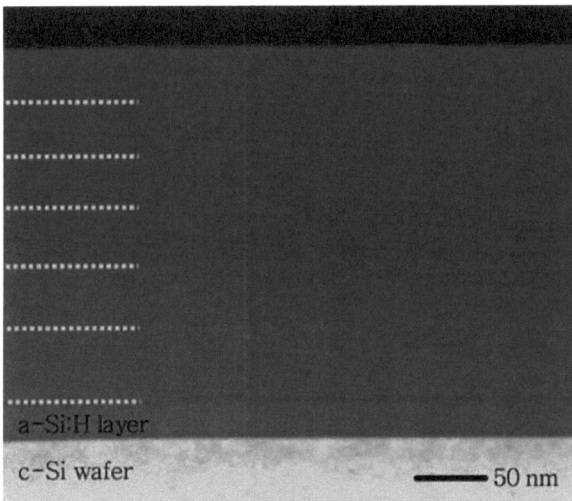

Figure 5.10: Cross-sectional TEM image of a 6-cycled pc-Si:H multilayer. Before the multilayer deposition, a 20 nm thick a-Si:H was deposited on a c-Si substrate. Adapted from S. W. Kwon, J. Kwak, S. Y. Myong, and K. S. Lim, Characterization of the protocrystalline silicon multilayer. *J. Non-Cryst. Solids*, **2006, 352, 1134. Copyright: Elsevier (2006).**

TEM Imaging of carbon nanomaterials

Common carbon allotropes include diamond, graphite, amorphrous C (a-C), fullerene (also known as buckyball), carbon nanotube (CNT, including single wall CNT and multi wall CNT), graphene. Most of them are chemically inert and have been found in nature. We can also define carbon as sp^2 carbon (which is graphite), sp^3 carbon (which is diamond) or hybrids of sp^2 and sp^3 carbon, as shown in Figure 5.11. As for carbon nanomaterials, fullerene, CNT and graphene are the three most well investigated, due to their unique properties in both mechanics and electronics. Under TEM, these carbon na-nomaterials will display three different projected images.

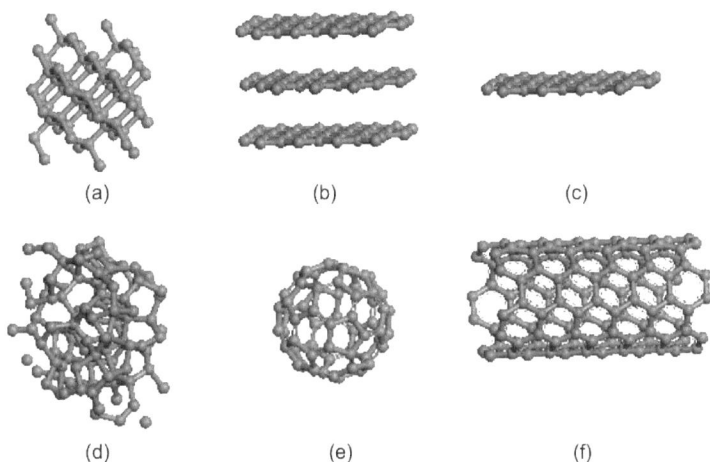

Figure 5.11: Six allotropes of carbon: a) diamond, b) graphite, c) graphene, d) amorphous carbon, e) C_{60} (Buckminsterfullerene or buckyball), f) single-wall carbon nanotube or buckytube.

All carbon naomaterials can be investigated under TEM. However, because of their difference in structure and shape, specific parts should be focused in order to obtain their atomic structure.

For C_{60}, which has a diameter of only 1 nm, it is relatively difficult to suspend a sample over a lacey carbon grid (a common kind of TEM grid usually used for nanoparticles). Even if the C_{60} sits on a thin a-C film, it also has some focus problems since the surface profile variation might be larger than 1 nm. One way to solve this problem is to encapsulate the C_{60} into single wall CNTs, which is known as nano peapods. This method has two benefits:

- CNT helps focus on C_{60}. Single wall is aligned in a long distance (rel-ative to C_{60}). Once it is suspended on lacey carbon film, it is much easier to focus on it. Therefore, the C_{60} inside can also be caught by minor focus changes.
- The CNT can protect C_{60} from electron irradiation. Intense high en-ergy electrons can permanently change the structure of the CNT. For C_{60}, which is more reactive than CNTs, it cannot survive after expos-ing to high dose fast electrons.

In studying CNT cages, C_{92} is observed as a small circle inside the walls of the CNT. While a majority of electron energy is absorbed by the CNT, the sample is still not irradiation-proof. Thus, as is seen in Figure 5.12, after a 123 s exposure, defects can be generated and two C_{92} fused into one new larger fullerene.

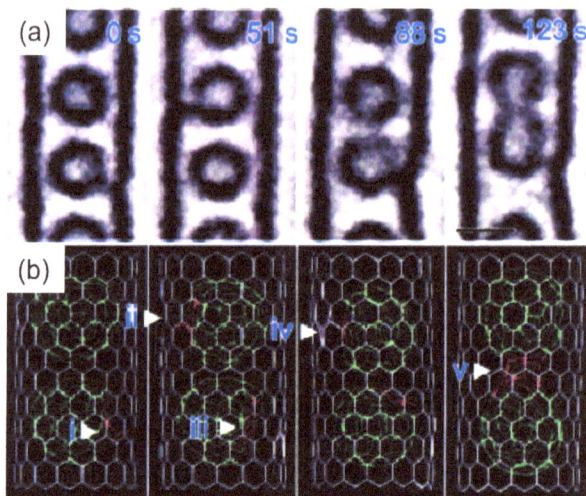

Figure 5.12: TEM images and calculated structures of C_{92} encapsulated in SWNTs under different electron irradiation time. Courtesy of Dr. Kazutomo Suenaga, adapted from K. Urita, Y. Sato, K. Suenaga, A. Gloter, A. Hasimoto, M. Ishida, T. Shimada, T. Shinohara, S. Iijima, Defect-induced atomic migration in carbon nanopeapod: tracking the single-atom dynamic behavior. *Nano Lett.*, 2004, 4, 2451. Copyright American Chemical Society (2004).

Although, the discovery of C_{60} was first confirmed by mass spectra rather than TEM. When it came to the discovery of CNTs, mass spectra were no longer useful because CNTs shows no individual peak in mass spectra since any sample contains a range of CNTs with different lengths and diameters. On the other hand, HRTEM can provide a clear image evidence of their existence. An example is shown in Figure 5.13.

Graphene is a planar fullerene sheet. Until recently, Raman, AFM and optical microscopy (graphene on 300 nm SiO_2 wafer) were the most convenient methods to characterize samples. However, in order to confirm graphene's atomic structure and determine the difference between monolayer and bilayer, TEM is still a good option. In Figure 5.14, a monolayer suspended graphene is observed with its atomic structure clearly shown. Inset is the FFT of the TEM image, which can be used as a filter to get an optimized structure image. High angle annular dark field (HAADF) image usually gives better contrast for different particles on it. It is also sensitive with changes of thickness, which allows a determination of the number of graphene layers.

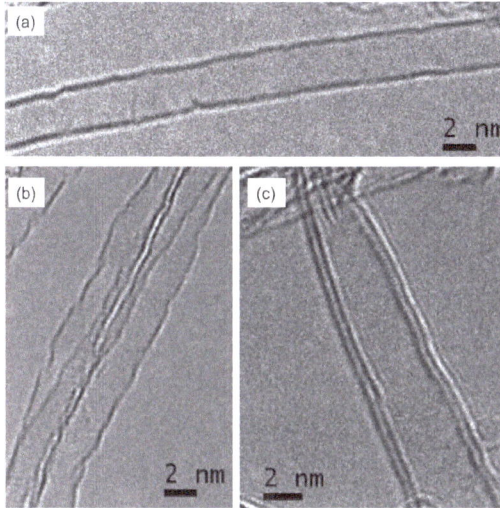

Figure 5.13: TEM images of SWNT and DWCNTs. Parallel dark lines corresponds to (002) lattice image of graphite. (a) and (b) SWNTs have 1 layer graphene sheet, diameter 3.2 nm. (c) DWCNT, diameter 4.0 nm.

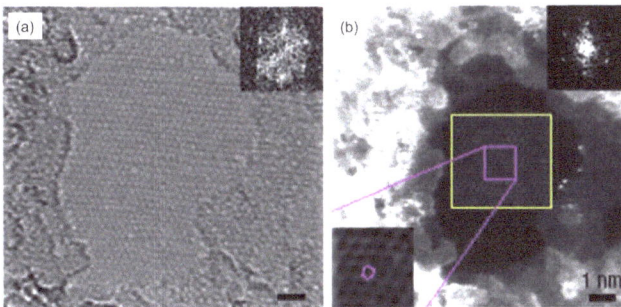

Figure 5.14: HRTEM of monolayer graphene. (a) Bright filed. (b) High Angle Annular Dark Field. Courtesy of M. H. Gass, adapted from M. H. Gass, U. Bangert, A. L. Bleloch, P. Wang, R. R. Nair, and A. K. Geim, Free-standing graphene at atomic resolution. *Nature Nanotechnol.*, 2008, 3, 676. Copyright: Nature Publishing Group (2008).

Graphene stacking and edges direction

Like the situation in CNT, TEM image is a projected image. Therefore, even with exact count of edge lines, it is not possible to conclude that a sample is a single layer graphene or multi-layer. If folding graphene has AA stacking (one layer is superposed on the other), with a projected direction of [001], one image could not tell the thickness of graphene. In order to distinguish such a bilayer of graphene from a single layer of graphene, a series of tilting experiment must be done. Different stacking structures of graphene are shown in Figure 5.15a.

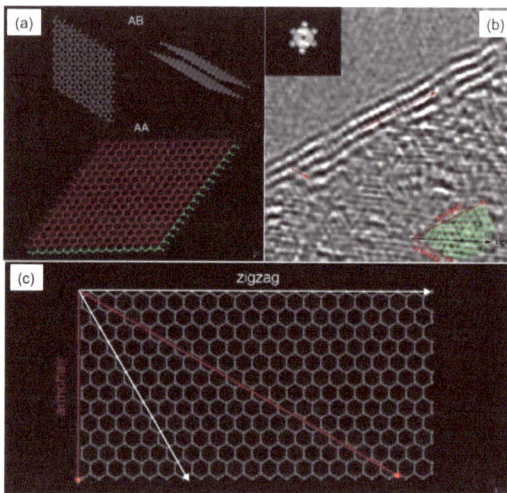

Figure 5.15: (a) Graphene stacking structure; (b) HRTEM image of graphene edges: zigzag and armchain (inset is FFT); (c) graphene edge model, a 30° angle between zigzag and armchair direction.

Theoretically, graphene has the potential for interesting edge effects. Based upon its sp^2 structure, its edge can be either that of a zigzag or armchair configuration. Each of these possess different electronic properties similar to that observed for CNTs. For both research and potential application, it is important to control the growth or cutting of graphene with one specific edge. But before testing its electronic properties, all the edges have to be identified, either by directly imaging with STM or by TEM. Detailed information of graphene edges can be obtained with HRTEM, simulated with fast Fourier transform (FFT). In Figure 5.15b, armchair directions are marked with red arrow respectively. A clear model in Figure 5.15c shows a 30° angle between zigzag edge and armchair edge.

TEM imaging of metal nanoparticles

Have you ever seen the beautiful Lycurgus Cup (Figure 5.16) in the movies or in the museum? Do you know why the Lycurgus Cup was red and could maintain the cranberry-like color for thousands of years? The answer is that it contains gold nanoparticles (NPs). The color of gold particles and their electronic structures vary as their size is changed, providing us with a new method to control the properties of metal nanoparticles. What is more, even if the number of atoms remains the same, their properties could differ significantly when they are made into different shapes like rods, disks and spheres. Thus, researchers want to know the microstructures of metal nanoparticles.

Figure 5.16: Lycurgus cup. Image by Lucas Livingston, 3 July, 2013, from https://www.flickr.com/photos/ancientartpodcast/9325599637.

Transmission electron microscopy (TEM), which shows the internal composition of our samples and gives a nanometer scale resolution, is an appropriate choice to characterize the structure of metal nanoparticles. In addition, the selected area electron diffraction (SAED) mode of TEM can provide more information about facets, phase and crystallinity. Those relatively large metal nanoparticles are usually not that sensitive to electron beams and thus could be observed with TEM following the normal procedure. However, some metal nanoparticles do require certain protective methods, like cryo-TEM, low-dose techniques and low-voltage TEM.

What information can be obtained from TEM images?

Morphology

TEM can be used to observe the microstructures of metal nanoparticles and sometimes can see individual atoms. Thus, it provides morphology information of metal nanoparticles, e.g., whether the particles are monodispersed or not, what is the average size and how their size distributes. Researchers synthesized triangular Au prisms with a seed-mediated growth method and then etched the prisms with $HAuCl_4$ and hexadecyltrimethylammonium bromide (CTAB, Figure 5.17), transforming the prisms into circular disks (Figure 5.18). TEM images not only showed the monodisperse of both precursors and products but also provided direct evidence for the etching process, the diameter varied <10% before and after etching. They deduced that the etching process only happened at the tips where the atoms had relatively lower coordination number.

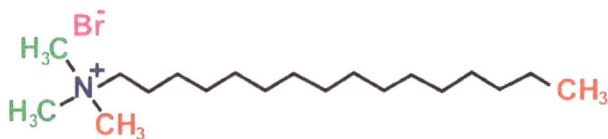

Figure 5.17: Structure of cetrimonium bromide [($C_{16}H_{33}$)N(CH_3)$_3$]Br (CTAB).

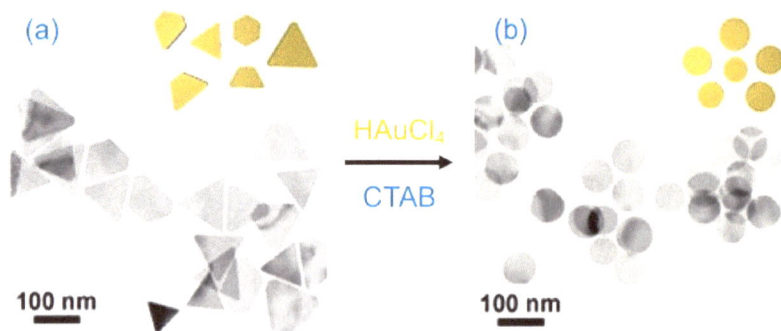

Figure 5.18: TEM image of Au nanodisks (a) before and after (b) etching. Adapted with permission from M. N. O'Brien, M. R. Jones, K. L. Kohlstedt, G. C. Schatz, and C. A. Mirkin, Uniform circular disks with synthetically tailorable diameters: two-dimensional nanoparticles for plasmonics. *Nano Lett.*, 2015, 15, 1012. Copyright: American Chemical Society (2015).

TEM can also be used to monitor the non-uniform deposition on the surface of nanoparticles. As is shown in Figure 5.19, researchers synthesized Al nanocrystals (d = 110 nm) and tried to deposit various transition metals on the surface. Instead of forming uniform layers, these transition metals tended to form islands (<10 nm in this case), which enables the decorated Al nanoparticles some special electronic and optical properties as well as excellent catalytic performance.

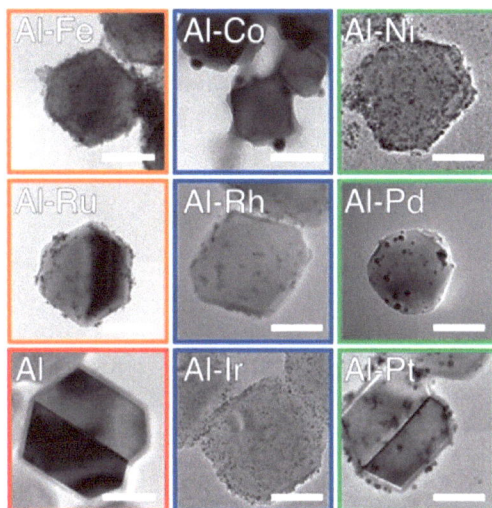

Figure 5.19: Al nanoparticles precursors and Al nanoparticles decorated with transition-metal islands. All scale bars are 50 nm. Adapted with permission from D. F. Swearer, R. K. Leary, R. Newell, S. Yazdi, H. Robatjazi, Y. Zhang, D. Renard, P. Nordlander, P. A. Midgley, N. J. Halas and E. Ringe, Transition-metal decorated aluminum nanocrystals. *ACS Nano*, 2017, 11, 10281. Copyright: American Chemical Society (2017).

Phase information

Phase information is mainly obtained under selected area electron diffraction (SAED) mode. TEM can switch to SAED mode by simply adjust the lenses and apertures. The selected area could be small and the corresponding diffraction pattern (DP) provides specific information about this area, e.g., facet, phase, crystallinity, and relates the surface information to the specific structure. Elongated Au tetrahexahedral (THH) crystals with seed-mediated growth. Since the 24 facets of THH crystals belong to the same crystal face, it is possible to reconstruct its structure by measuring the angles between the facets and various crystal planes in TEM images (Figure 5.20). Tilting the elongated THH Au crystal to its [100] projection, confirmed by SAED, and measuring the angle between the bevels and bases (the angle labeled in red in Figure 5.20) and then tilted the crystal a little bit, confirmed it was [310] projection. Repeating this process resulted in enough data. The angles obtained were close to the angle between {037} facets and corresponding facets. This result was further confirmed with high-resolution TEM (HRTEM) results and modeling. Therefore, the facets of elongated THH Au nanoparticles were determined.

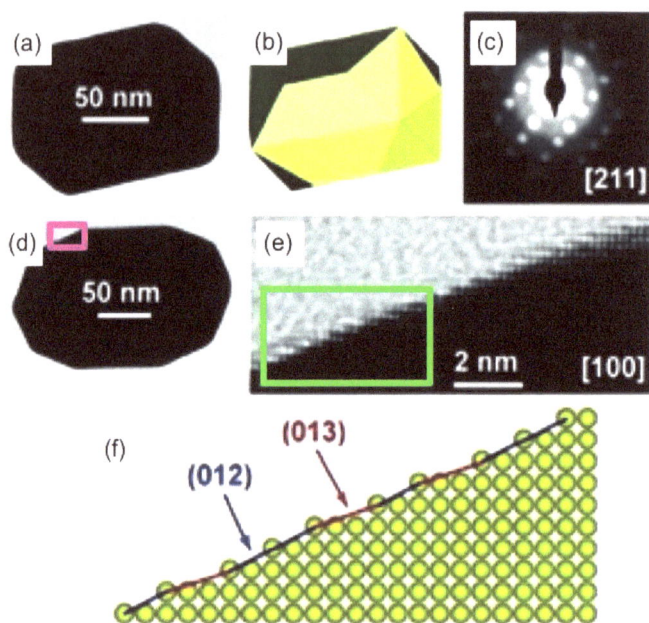

Figure 5.20: Elongated Au tetrahexahedral (THH) crystals with seed-mediated growth: (a and d) TEM images of elongated Au THH nanoparticles from different directions; (b) schematic model from corresponding directions; (c) SAED pattern of the central part of elongated Au THH nanoparticles from the same directions as in TEM images, (e) a high-resolution TEM (HRTEM) of the region inside the purple box in (d), and (f) atomic model of the region inside the green box in (k). Adapted with permission from T. Ming, W. Feng, Q. Tang, F. Wang, L. Sun, J. Wang, and C. Yan, Growth of tetrahexahedral gold nanocrystals with high-index facets. *J. Am. Chem. Soc.*, 2009, 131, 16350. Copyright: American Chemical Society (2009).

Assembly

Metal nanoparticles are not simply isolated from the outer environment; they could interact with ligand molecules, solvent molecules and even other nanoparticles. Thus, with proper design, metal nanoparticles could be assembled and form certain structures. One example is when Fe_2O_3 and Au nanoparticles with proper diameters were mixed in solution and 'crystalized' with the solvent vaporized, like what happened to ions with proper diameters, these nanocrystals could form NaCl-type binary superlattices (Figure 5.21). This could also be applied to other nanoparticles systems e.g., PbSe and Pd nanoparticles, Au and Ag nanoparticles, Fe and Au nanoparticles.

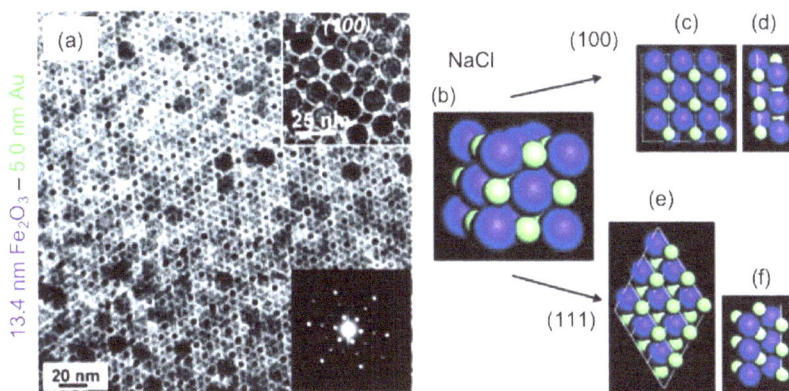

Figure 5.21: TEM images of NaCl-type nanoparticles: (a) (111) projection built from 13.4 nm Fe_2O_3 and 5.0 nm Au nanoparticles. Up insert shows the (100) projection. Bottom insert is small-angle ED pattern; (b) 3D structure of NaCl unit cell; (c and d) different crystal facets of NaCl; (e and f) side view of (e and g). Adapted from E. V. Shevchenko, D. V. Talapin, C. B. Murray, and S. O'Brien, Structural characterization of self-assembled multifunctional binary nanoparticle superlattices. *J. Am. Chem. Soc.*, 2006, 128, 3620. Copyright: American Chemical Society (2006).

Advances in TEM techniques related to metal NPs

Liquid-cell TEM

Although TEM is operated under high vacuum and liquid samples cannot be loaded into TEM directly, it has been one of the ultimate goals to image liquid samples the same as solid samples in TEM since the invention of TEM. This technology is critical for metal nanoparticles because: most metal nanoparticles are synthesized and dispersed in solutions and their shape and behavior in solution are different from that on the substrate or in the vacuum. In situ monitoring helps us understand how metal nanoparticles behavior in liquid and gives us more insights.

There had been some process in imaging liquids with low vapor pressure or encapsulated by another material. However, most types of solvents like water and ethanol require special equipment. For example, in order to contain the samples, an earlier system used an open environmental chamber where the pressure could maintain up to 0.2 bar. The other part of TEM remains high vacuum with the help of differential pumping (Figure 5.22a). This system has been used in today's *in situ* TEM while it is difficult to control the thickness of the sample to <1 μm, as is required to obtain nanometer-scale resolution. Thus, open environmental chambers are not suitable for HRTEM. Besides, it is also not applicable when flowing liquid is needed. A closed cell with a certain thickness is our certain choice to solve this problem, as is shown in Figure 5.22b. Based on this idea various systems were developed in the early years, although the spatial resolution is not much better than VLMs that are much easier to use.

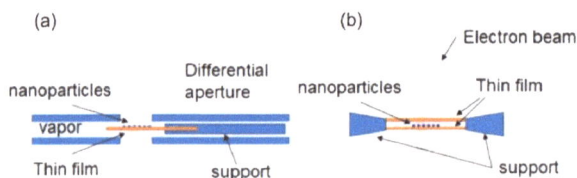

Figure 5.22: Schematic representation of (a) the structure of an environmental chamber in TEM, which is isolated from the other part of TEM by differential apertures and (b) the structure of a sample holder which enclose nanoparticles and liquid between "windows"

The rapid development of thin-film technology, microfabrication and imaging in the 2000s brought a breakthrough to liquid cells. Thin carbon foils and graphene sheets have been used as membranes, which helped the *in situ* imaging of metal nanoparticles. For example, researchers used 2 sheets of graphene as the liquid cell and study the movement of nanoparticle dimers connected with dsDNA (Figure 5.23). They found that: the dimers could fluctuate in their distance. Sometimes the nanoparticles even overlap and then separate from each other slightly, indicating the existence of 3D rotation in this dimer system. Besides, this sample also shows a weak substrate-nanoparticle attraction, which may perturb 3D dynamics of such systems.

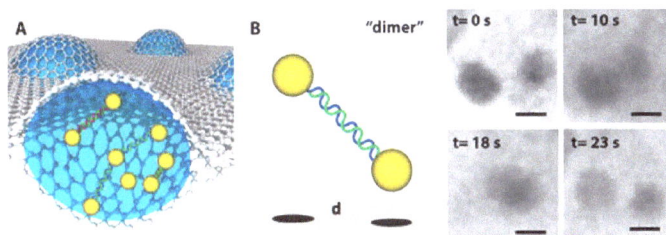

Figure 5.23: (a) A schematic of a graphene liquid cell encapsulating single nanoparticles, dimers and trimers connected with dsDNA, (b) Schematic of a dimer and its projection onto the screen of TEM camera, and (c) TEM images of a rotating dimer. Adapted from Q. Chen, J. M. Smith, J. Park, K. Kim, D. Ho, H. I. Rasool, A. Zettl and A. P. Alivisatos, 3D motion of DNA-Au nanoconjugates in graphene liquid cell electron microscopy. *Nano Lett.*, 2013, 13, 4556. Copyright: American Chemical Society (2013).

Despite the success of graphene sheets, liquid cells made from thin Si_3N_4 films (20-100 nm) and silicon microchips are increasingly popular since they are robust, easy to manufacture and homogenous in composition and thickness. Combining silicon processing techniques, it is possible to control the temperature of the sample and to perform electrochemical tests and inject liquids into the cells. A typical liquid cell for in-situ electrochemical test is shown in Figure 5.24. Ideally, the thinner the films are, the better the resolution will be. However, thin films will bow under vacuum and increase the thickness of liquid in the center. Therefore, balancing the thickness of films used, window size and window thickness is very critical for TEM experiments in liquid cells. Besides, bubbles in the cell caused by electron beams may also cause problems.

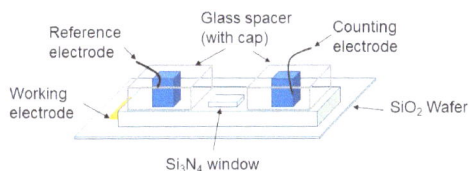

Figure 5.24: Schematic of a liquid cell made from Si microchips and Si_3N_4 windows along with an electrochemical test system.

3D reconstruction of metal nanoparticles with TEM

Electron tomography is widely used in the 3D analysis of various materials. Unfortunately, limited by its mechanism (imaging static materials from different tilt angles for 2 to 5 hours), it cannot be used in the characterization of metal nanoparticles and could be damaged due to high doses high-energy electrons, especially those which are extremely small and active. However, reconstruction of metal nanoparticles is important to illustrate the formation mechanism, stability and the origin of their catalytic performance. Therefore, 3D reconstruction of metal nanoparticles has been developed in recent years and various methods have been developed.

Considering the problems of electron tomography for metal nanoparticles, researchers used a large number of nanoparticles (~4300 Au_{144} nanoparticles) to reduce the doses or each nanoparticle and to obtain structure information from different directions. As-obtained images were classified into 200 classes and averaged to generate the initial structure. A projection matching approach was used to give the final structure. The reconstruction is shown in Figure 5.25. The positions of most atoms are in accord with face-centered cubic (FCC) packing while some deviate from ideal position to confer the curvature of the nanoparticle.

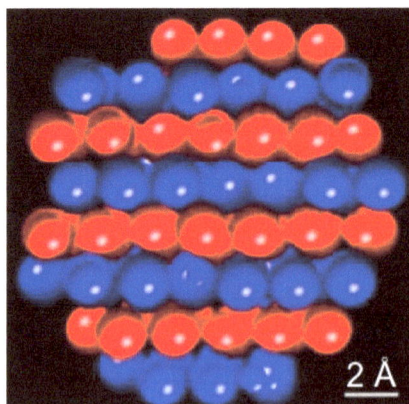

Figure 5.25: Reconstruction of the central section of an Au_{144} particle showing atoms packed with fcc symmetry. Adapted from M. Azubel, A. L. Koh, K. Koyasu, T. Tsukuda and R. D. Kornberg, Structure determination of a water-soluble 144-gold atom particle at atomic resolution by aberration-corrected electron microscopy. *ACS Nano*, 2017, 11, 11866. Copyright: American Chemical Society (2017).

Transmission electron energy loss spectroscopy

Electron energy loss spectroscopy (EELS) is a technique that measures electronic excitations within solid-state materials. When an electron beam with a narrow range of kinetic energy is directed at a material some electrons will be inelastically scattered, resulting in a kinetic energy loss. Electrons can be inelastically scattered from phonon excitations, plasmon excitations, interband transitions, or inner shell ionization. EELS measures the energy loss of these inelastically scattered electrons and can yield information on atomic composition, bonding, electronic properties of valance and conduction bands and surface properties. An example of atomic level composition mapping is shown in Figure 5.26a. EELS has even been used to measure pressure and temperature within materials.

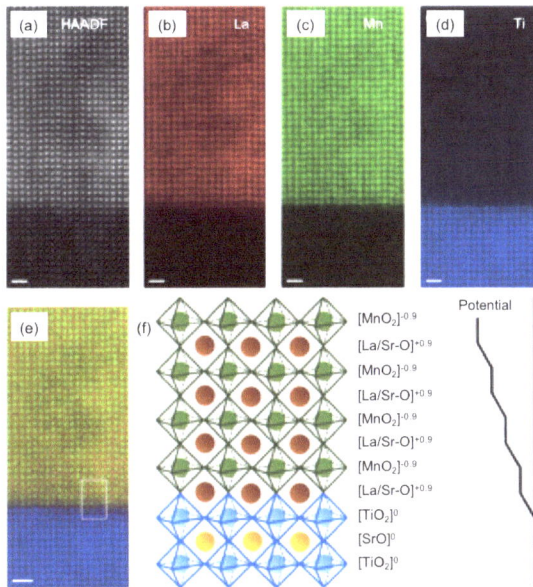

Figure 5.26: EEL Map showing atomic composition of a $La_xSr_{1-x}MnO_3$ film grow on TiO_2 with atomic scale spatial resolution: (a) high angle annular dark field (HAADF) image, (b) La (c) Mn (d) Ti elemental data obtain from a STEM-EELS measurement, (e) overlayed image of b-d, and (f) model of theoretical packing in $La_xSr_{1-x}MnO_3$ film. Adapted from J. A. Mundy, Y. Hikita, T. Hidaka, T. Yajima, T. Higuchi, H. Y. Hwang, D. A. Muller, and L. F. Kourkoutis, Visualizing the interfacial evolution from charge compensation to metallic screening across the manganite metal–insulator transition. *Nat. Commun.*, 2014, 5, 3464. Copyright Nature Publishing Group 2015.

The EEL spectrum

An idealized EEL spectrum is show in Figure 5.27. The most prominent feature of any EEL spectrum is the zero-loss peak (ZLP). The ZLP is due to those electrons from the electron beam that do not inelastically scatter and reach the detector with their original kinetic energy; typically, 100 - 300 keV. By definition the ZLP is set to 0 eV for further analysis and all signals arising from inelastically scatter electrons occur at >0 eV. The second largest feature is often the plasmon resonance: the collective excitation of conduction band electrons within a material. The plasmon resonance and other peaks attributed to weakly bound, or outer shell electrons, occur in the "low-loss" region of the spectrum. The low-loss regime is typically thought of as energy loss <50 eV, but this cut-off from low-loss to high-loss is arbitrary. Shown in the inset of Figure 5.27 is an edge from atom core-loss and further fine structure. Inner shell ionizations, represented by the core-loss peaks, are useful in determining elemental compositions as these peaks can act as fingerprints for specific elements. For example, if there is a peak at 532 eV in an EEL spectrum, there is a high probability that the sample contains a considerable amount of oxygen because this is known to be the energy needed to remove an inner shell electron from oxygen. This idea is further explored by looking at sudden changes in the bulk plasmon for aluminum in different chemical environments as shown in Figure 5.28.

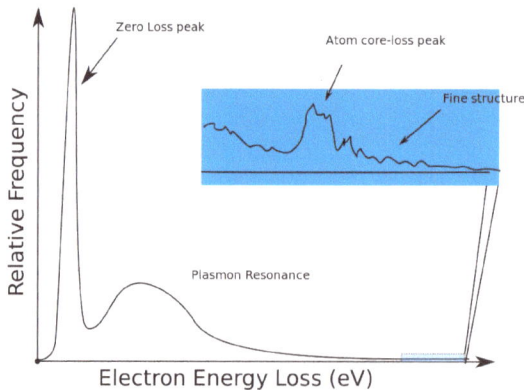

Figure 5.27: Idealized electron energy loss spectrum (EELS). Inset shows core loss and fine structure. This image is filed under the Creative Commons Attribution-Share Alike 3.0 Unported License. Original Author Hat'n'Coat.

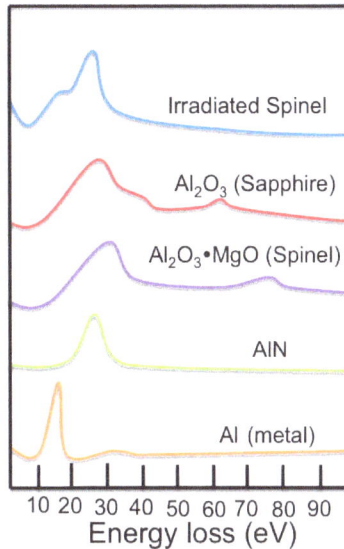

Figure 5.28: Slight shifts in the plasmon peak of pure Al, AlN, and various aluminum oxides. Slight shifts in peak shape and energy can allow EEL spectroscopists to determine slight variations in chemical and electron enviroments. Adapted from D. B. William and C. B. Carter. *Transmission Electron Microscopy: A Textbook for Materials Science.* **2ⁿᵈ edn., Springer, New York, NY, (2009).**

Of course, there are several other techniques available for probing atomic compositions many of which are covered in this text. These include Energy dispersive X-ray spectroscopy, X-ray photoelectron spectroscopy, and Auger electron spectroscopy. Please reference these chapters for a thorough introduction to these techniques.

Electron energy loss spectroscopy versus energy dispersive X-ray spectroscopy

As a technique EELS is most frequently compared to energy dispersive X-ray spectroscopy (EDX) also known as energy dispersive spectroscopy (EDS). Energy dispersive X-ray detectors are commonly found as analytical probes on both scanning and transmission electron microscopes. The popularity of EDS can be understood by recognizing the simplicity of compositional analysis using this technique. However, EELS data can offer complementary compositional analysis while also generally yielding further insight into the solid-state physics and chemistry in a system at the cost of a steeper learning curve. EDS and EELS spectra are both derived from the electronic excitations of materials, however, EELS probes the initial excitation, while EDS looks at X-ray emissions from the decay of this excited state. As a result, EEL spectra investigate energy ranges from 0-3 keV while EDS spectra analyze a wider energy range from 1-40 keV. The difference in ranges makes EDS suited particularly well for heavy elements while EELS complements measurement elements lighter than Zn.

History and Implementation

In the early 1940s, James Hillier (Figure 5.29) and R. F. Baker were looking to develop a method for pairing the size, shape, and structure available from electron microscopes to a convenient method for "determining the composition of individual particles in a mixed specimen". Their instrument reported in 1944 was the first electron-optical instrument used to measure the velocity distribution in an electron beam transmitting through a sample.

Figure 5.3290: Canadian scientist and inventor James Hillier (1915 - 2007) who is recognized as a pioneer in the field of electron microscopy.

The instrument was built from a repurposed transmission electron microscope (TEM). It consisted of an electron source and three electromagnetic focusing lenses, standard for TEMs at the time, but also incorporated a magnetic deflecting lenses, which when turned on, would redirect the electrons 180° into a photographic plate. The electrons with varying kinetic energies dispersed across the photographic plate and could be correlated to the energy loss of each peak depending on position. In this groundbreaking work, Hillier and Baker were able to find the discrete energy loss corresponding to the K levels of both carbon and oxygen.

The vast majority of EEL spectrometers are found as secondary analyzers in transmission electron microscopes. It wasn't until the 1990s when EELS became a widely used research tool because of advances in electron beam aberration correction and vacuum technologies. Today, EELS is capable of spatial resolutions down to the single atom level, and if the electron beam is monochromated then energy resolution can be as low as 0.01eV. Figure 5.30 depicts the typical layout of an EEL spectrometer at the base of a TEM.

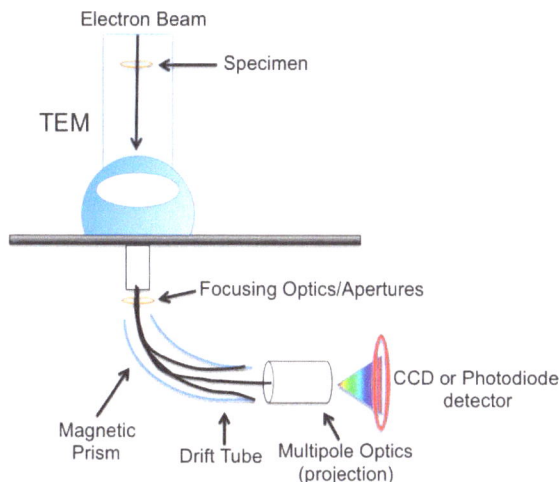

Figure 5.30: Schematic to show the simplified positions of standard EELS components in a standard TEM.

Experimental characterization of metal nanoparticles by TEM

Usually, the images are saved with CCD detectors, although some older TEMs still use film to record the signals. As-collected data could then be analyzed with powerful software such as Digital Micrograph.

- **Sample solution preparation.** We will first dilute the solution of nanoparticles with solvent (Milli-Q water is usually used) and treat it with sonication. Make sure the samples are well dispersed without aggregation.
- **E-chip preparation.** Here we take an E-chip produced by Protochips Inc. as an example. Before use, the protective layer should be removed by rinsing the liquid cell in acetone and ethanol solvents

(about 2 min for each). Later on, the E-chip is cleaned with plasma cleaning (9:1 argon/oxygen mixture).

- **Chip loading.** As-prepared E-chip was loaded onto a Poseidon 210 liquid fluid holder and fixed firmly with screws.
- **Preparation of TEM.** The vacuum system should be turned on in advance. For some TEM instrument like JEOL 2100F, we have to add liquid nitrogen into the Dewar so as to keep the cold fingers cold.
- **Insert the samples.** The sample holder is inserted into the TEM after the pump of the sample chamber was turned on. After the light turns green, turn the holder counterclockwise and hold it as the vacuum sucks the holder in.
- **Injection of nanoparticle solution.** The solution is injected into the sample holder with a syringe pump.
- **Turn on electron beam.**
- **Alignment.** Move the electron beam to the center of the phosphor screen. Choose the right condense lenses and center the beam by rotating the dials in the lens. Adjust sample height and stigmation. Repeat this step several times.
- **Focus image.** Find your samples at low magnification, increase the magnification and then condense the beam. Adjust the position of the objective lens until the diffraction rings disappear. Increase the magnification and repeat this process again.
- **Take pictures.** Adjust the brightness to a proper value and find your samples. Set up the camera and open the screen. Take pictures.
- **Shut down.** Close the screen, turn off the beam and the camera, take out the sample.

Bibliography

M. Azubel, A. L. Koh, K. Koyasu, T. Tsukuda and R. D. Kornberg, Structure determination of a water-soluble 144-gold atom particle at atomic resolution by aberration-corrected electron microscopy. *ACS Nano*, 2017, **11**, 11866.

J. Campos-Delgado, J. M. Romo-Herrera, X. Jia, D. A. Cullen, H. Muramatsu, Y. A. Kim, T. Hayashi, Z. Ren, D. J. Smith, Y. Okuno, T. Ohba, H. Kanoh, K. Kaneko, M. Endo, H. Terrones, M. S. Dresselhaus, and M. Terrones, *Nano Lett.*, 2008, **8**, 2773.

Q. Chen, J. M. Smith, J. Park, K. Kim, D. Ho, H. I. Rasool, A. Zettl and A. P. Alivisatos, 3D motion of DNA-Au nanoconjugates in graphene liquid cell electron microscopy. *Nano Lett.*, 2013, **13**, 4556.

N. de Jonge, M. Pfaff and Peckys, D. B., *Practical aspects of transmission electron microscopy in liquid. In Advances in Imaging and Electron Physics*, ed. P. W. Hawkes, Elsevier, UK, 1st edn., Vol. 186, pp. 1-37.

N. de Jonge and F. M. Ross, Electron microscopy of specimens in liquid. *Nat. Nanotechnol.*, 2011, **6**, 695.

R. F. Egerton, Control of radiation damage in the TEM. *Ultramicroscopy*, 2013, **127**, 100.

S. Eustis and M. A. El-Sayed, Why gold nanoparticles are more precious than pretty gold: Noble metal surface plasmon resonance and its enhancement of the radiative and nonradiative properties of nanocrystals of different shapes. *Chem. Soc. Rev.*, 2006, **35**, 209.

M. H. Gass, U. Bangert, A. L. Bleloch, P. Wang, R. R. Nair, and A. K. Geim, Free-standing graphene at atomic resolution. Free-standing graphene at atomic resolution. *Nature Nanotechnol.*, 2008, **3**, 676.

J. Hillier and R. F. Baker, *J. Appl. Phys.*, 1944, **15**, 663.

S. W. Kwon, J. Kwak, S. Y. Myong, and K. S. Lim, Characterization of the protocrystalline silicon multilayer. *J. Non-Cryst. Solids*, 2006, **352**, 1134.

T. Ming, W. Feng, Q. Tang, F. Wang, L. Sun, J. Wang, and C. Yan, Growth of tetrahexahedral gold nanocrystals with high-index facets. *J. Am. Chem. Soc.*, 2009, **131**, 16350.

J. A. Mundy, Y. Hikita, T. Hidaka, T. Yajima, T. Higuchi, H. Y. Hwang, D. A. Muller, and L. F. Kourkoutis, Visualizing the interfacial evolution from charge compensation to metallic screening across the manganite metal–insulator transition. *Nat. Commun.*, 2014, **5**, 3464.

K. W. Noh, Y. Liu, L. Sun, and S. J. Dillon, Challenges associated with in-situ TEM in environmental systems: The case of silver in aqueous solutions. *Ultramicroscopy*, 2012, **116**, 34.

M. N. O'Brien, M. R. Jones, K. L. Kohlstedt, G. C. Schatz, and C. A. Mirkin, Uniform circular disks with synthetically tailorable diameters: two-dimensional nanoparticles for plasmonics. *Nano Lett.*, 2015, **15**, 1012.

J. Park, H. Elmlund, P. Ercius, J. M. Yuk, D. T. Limmer, Q. Chen, K. Kim, S. H. Han, D. A. Weitz, A. Zettl and A. P. Alivisatos, 3D structure of individual nanocrystals in solution by electron microscopy. *Science*, 2015, **349**, 290.

Q. M. Ramasse, C. R. Seabourne, D.-M. Kepaptsoglou, R. Zan, U. Bangert, and A. J. Scott, Probing the bonding and electronic structure of single atom dopants in graphene with electron energy loss spectroscopy. *Nano Lett.*, 2013, **13**, 4989.

L. Reimer and H. Kohl, *Transmission Electron Microscopy Physics of Image Formation*, 5th edn., Springer, New York (2008).

E. V. Shevchenko, D. V. Talapin, C. B. Murray, and S. O'Brien, Structural characterization of self-assembled multifunctional binary nanoparticle superlattices. *J. Am. Chem. Soc.*, 2006, **128**, 3620.

D. Shindo and K. Hiraga, *High-Resolution Electron Microscopy for Material Science*, Springer, New York (1998).

D. L. Spector and R. D. Goldman, *Basic Methods in Microscopy Protocols and Concepts from Cells: A Laboratory Manual*, Cold Spring Harbor, New York (2006).

D. F. Swearer, R. K. Leary, R. Newell, S. Yazdi, H. Robatjazi, Y. Zhang, D. Renard, P. Nordlander, P. A. Midgley, N. J. Halas and E. Ringe, Transition-metal decorated aluminum nanocrystals. *ACS Nano*, 2017, **11**, 10281.

K. Urita, Y. Sato, K. Suenaga, A. Gloter, A. Hasimoto, M. Ishida, T. Shimada, T. Shinohara, S. Iijima, Defect-induced atomic migration in carbon nanopeapod: tracking the single-atom dynamic behavior. *Nano Lett.*, 2004, **4**, 2451.

D. B. Williams and C. B. Carter, *Transmission Electron Microscopy: A Textbook for Material Science*, 2nd edn., Springer, New York (2009).

M. J. Williamson, R. M. Tromp, P. M. Vereecken, R. Hull, and F. M. Ross, Dynamic microscopy of nanoscale cluster growth at the solid–liquid interface. *Nat. Mater.*, 2003, **2**, 532.

D. L. Windt, F. E. Christensen, W. W. Craig, C. Hailey, F. A. Harrison, M. Jimenez-Garate, R. Kalyanaraman, and P. H. Mao, Growth, structure, and performance of depth-graded W/Si multilayers for hard X-ray optics. *J. Appl. Phys.*, 2000, **88**, 460.

Y. Yabuuchi, S. Tametou, T. Okano, S. Inazato, S. Sadayamn, and Y. Tamamoto, A study of the damage on FIB-prepared TEM samples of $Al_xGa_{1-x}As$. *J. Electron Microsc.*, 2004, **53**, 5.

Z. Zeng, W. Zheng, and H. Zheng, Visualization of colloidal nanocrystal formation and electrode–electrolyte interfaces in liquids using TEM. *Acc. Chem. Res.*, 2017, **50**, 1808.

L. Zheng and S. Iijima, Free folding of suspended graphene sheets by random mechanical stimulation. *Phys. Rev. Lett.*, 2009, **102**, 015501.

Chapter 6: Scanning Tunneling Microscopy

Michelle LaComb, Zhiwei Peng, Muqing Ren, Pavan M. V. Raja and
Andrew R. Barron

Introduction

Scanning tunneling microscopy (STM) is a powerful instrument that allows one to image the sample surface at the atomic level. As the first generation of scanning probe microscopy (SPM), STM paves the way for the study of nano science and nano-materials. For the first time, researchers could obtain atom-resolution images of electrically conductive surfaces as well as their local electric structures. Because of this milestone invention, Gerd Binnig (Figure 6.1) and Heinrich Rohrer (Figure 6.2) were awarded the Nobel Prize in Physics in 1986.

Figure 6.1: German physicist Gerd Binnig (1947 -).

Figure 6.2: Swiss physicist Heinrich Rohrer (1933 -)

Principles of scanning tunneling microscopy

The key physical principle behind STM is the *tunneling effect*. In terms of their wave nature, the electrons in the surface atoms actually are not as tightly bonded to the nucleons as the electrons in the atoms of the bulk. More specifically, the electron density is not zero in the space outside the surface, though it will decrease exponentially as the distance between the electron and the surface increases (Figure 6.3a). So, when a metal tip approaches to a conductive surface within a very short distance, normally just a few Å, their perspective electron clouds will start to overlap, and generate tunneling current if a small voltage is applied between them, as shown in Figure 6.3b.

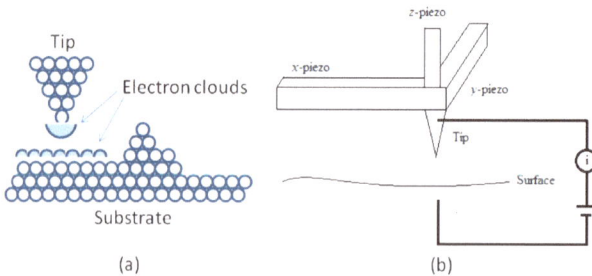

(a)   (b)

Figure 6.3: Schematic diagram of the principles of AFM showing (a) the interactions between tip and surface and (b) the tunneling current generated from tip and surface is measured and used as feedback to control the movement of the tip.

When we consider the separation between the tip and the surface as an ideal one-dimensional tunneling barrier, the tunneling probability, or the tunneling current I, will depend largely on s, the distance between the tip and surface,

$$I \propto \exp(-2s \,|2m/h^2 \,(\langle \phi \rangle - e\,|V\,|/2)|^{1/2}$$

where m is the electron mass, e the electron charge, h the Plank constant, ϕ the averaged work function of the tip and the sample, and V the bias voltage. A simple calculation will show us how strongly the tunneling current is affected by the distance (s). If s is increased by $\Delta s = 1$ Å,

$$\Delta I = \exp(-2k_0 \,\Delta s)$$

$$k_0 = |2m/h^2 \,(\langle \phi \rangle - e\,|V\,|/2)|^{1/2}$$

Usually ($<\phi>$ -e$|$V$|$/2) is about 5 eV, which k_0 about 1 Å^{-1}, then $\Delta I/I = {}^1/_8$. That means, if s changes by 1 Å, the current will change by one order of the magnitude. That's the reason why we can get atom-level image by measuring the tunneling current between the tip and the sample.

In a typical STM operation process, the tip is scanning across the surface of sample in x-y plain, the instrument records the x-y position of the tip, measures the tunneling current, and control the height of the tip via a feedback circuit. The movements of the tip in x, y and z directions are all controlled by piezo ceramics, which can be elongated or shortened according to the voltage applied on them.

Normally, there are two modes of operation for STM, *constant height mode* and *constant current mode*. In constant height mode, the tip stays at a constant height when it scans through the sample, and the tunneling current is measured at different (x, y) position (Figure 6.3b). This mode can be applied when the surface of sample is very smooth. But, if the sample is rough, or has some large particles on the surface, the tip may contact, with the sample and damage the surface. In this case, the constant current mode is applied. During this scanning process, the tunneling current, namely the distance between the tip and the sample, is settled to an unchanged target value. If the tunneling current is higher than that target value, that means the height of the sample surface is increasing, the distance between the tip and sample is decreasing. In this situation, the feedback control system will respond quickly and retract the tip. Conversely, if the tunneling current drops below the target value, the feedback control will have the tip closer to the surface. According to the output signal from feedback control, the surface of the sample can be imaged.

The main component of a scanning tunneling microscope is a rigid metallic probe tip, typically composed of tungsten, connected to a piezodrive containing three perpendicular piezoelectric transducers (Figure 6.4). The tip is brought within a fraction of a nanometer of an electrically conducting sample. At close distances, the electron clouds of the metal tip overlap with the electron clouds of the surface atoms (Figure 6.4, inset). If a small voltage is applied between the tip and the sample a tunneling current is generated. The magnitude of this tunneling current is dependent on the bias voltage applied and the distance between the tip and the surface. A current amplifier can covert the generated tunneling current into a voltage. The magnitude of the resulting voltage as compared to the initial voltage can then be used to control the piezodrive, which controls the distance between the tip and the surface (i.e., the z direction). By scanning the tip in the x and y directions, the tunneling current can be measured across the entire sample. The STM system can operate in either of two modes:

- Constant height.
- Constant current.

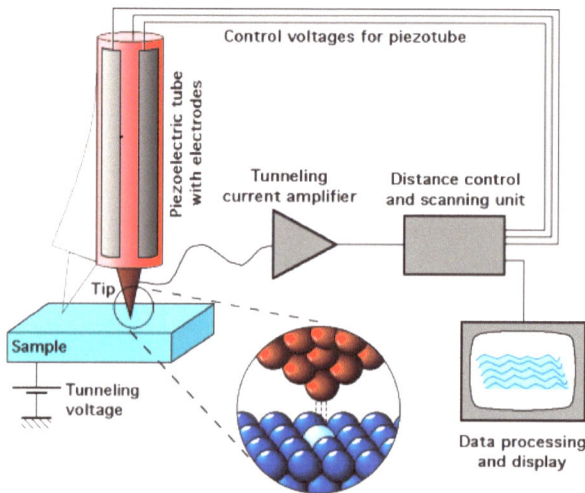

Figure 6.4: Schematic drawing of a STM apparatus.

In constant height mode, the tip is fixed in the z direction and the change in tunneling current as the tip changes in the x,y direction is collected and plotted to describe the change in topography of the sample. This method is dangerous for use in samples with fluctuations in height as the fixed tip might contact and destroy raised areas of the sample. A common method for non-uniformly smooth samples is constant current mode. In this mode, a target current value, called the set point, is selected and the tunneling current data gathered from the sample is compared to the target value. If the collected voltage deviates from the set point, the tip is moved in the z direction and the voltage is measured again until the target voltage is reached. The change in the z direction required to reach the set point is recorded across the entire sample and plotted as a representation of the topography of the sample. The height data is typically displayed as a gray scale image of the topography of the sample, where lighter areas typically indicate raised sample areas and darker spots indicate protrusions. These images are typically colored for better contrast.

The standard method of STM, described above, is useful for many substances (including high precision optical components, disk drive surfaces, and Buckyballs) and is typically used under ultrahigh vacuum to avoid contamination of the samples from the surrounding systems. Other sample types, such as semiconductor interfaces or biological samples, need some enhancements to the traditional STM apparatus to yield more detailed sample information. Three such modifications, spin-polarized STM (SP-STM), ballistic electron emission microscopy (BEEM) and photon STM (PSTM) are summarized in Table 6.1 and in described in detail below.

Name	Alterations to conventional STM	Sample types	Limitations
STM	None	Conducting surface	Rigidity of probe
SP-STM	Magnetized STM tip	Magnetic	Needs to be overlaid with STM, magnetized tip type
BEEM	Three-terminal with base electrode and current collector	Interfaces	Voltage, changes due to barrier height
PSTM	Optical fiber tip	Biological	Optical tip and prism manufacture

Table 6.1: Comparison of conventional STM and alterations.

Comparison of atomic force microscopy and scanning tunneling microscopy

Both AFM and STM are widely used in nano science. According to the different working principles though, they have their own advantages and disadvantages when measuring specific properties of sample (Table 6.2). STM requires an electric circuit including the tip and sample to let the tunneling current go through. That means, the sample for STM must be conducting. In case of AFM however, it just measures the deflection of the cantilever caused by the van der Waals forces between the tip and sample. Thus, in general any kind of sample can be used for AFM. But, because of the exponential relation of the tunneling current and distance, STM has a better resolution than AFM. In STM image one can actually "see" an individual atom, while in AFM it's almost impossible, and the quality of AFM image is largely depended on the shape and contact force of the tip. In some cases, the measured signal would be rather complicated to interpret into morphology or other properties of sample. On the other side, STM can give straight forward electric property of the sample surface.

	AFM	STM
Sample requirement	-	Conducting
Work environment	Air, liquid	Vacuum
Lateral resolution	~1 nm	~0.1 nm
Vertical resolution	~0.05 nm	~0.05 nm
Working mode	Tapping, contact	Constant current, constant height

Table 6.2: Comparison of AFM and STM.

Applications of scanning tunneling microscopy in nano-science

STM provides a powerful method to detect the surface of conducting and semi-conducting materials. Recently STM can also be applied in the imaging of insulators, superlattice assemblies and even the manipulation of molecules on surface. More importantly, STM can provide the surface structure and electric property of surface at atomic resolution, a true breakthrough in the development of nano science. In this sense, the data collected from STM could reflect the local properties even of single molecule and atom. With these valuable measurement data, one could give a deeper understanding of structure-property relations in nanomaterials.

An excellent example is the STM imaging of graphene on Ru(0001), as shown in Figure 6.5. Clearly seen is the superstructure with a periodicity of ~30 Å, coming from the lattice mismatch of 12 unit cells of the graphene and 11 unit cells of the underneath Ru(0001) substrate. This so-called moiré structure can also be seen in other systems when the adsorbed layers have strong chemical bonds within the layer and weak interaction with the underlying surface. In this case, the periodic superstructure seen in graphene tells us that the formed graphene is well crystallized and expected to have high quality.

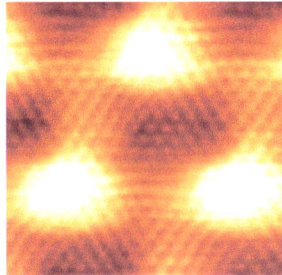

Figure 6.5: Atomically resolved image of the graphene overlayer. The scanning area is 40×40 Å, the operation mode is constant current mode. Adapted with permission from S. Marchini, S. Gunther, and J. Wintterlin, Scanning tunneling microscopy of graphene on Ru(0001). *Phys. Rev. B*, 2007, 76, 075429. Copyrighted by the American Physical Society.

Another good example is shown to see that the measurement from STM could tell us the bonding information in single-molecular level. In thiol- and thiophene-functionalization of single-wall carbon nanotubes (SWNTs, Figure 6.6), the use of Au nanoparticles as chemical markers for AFM gives misleading results, while STM imaging could give correct information of substituent location. From AFM image, Au-thiol-SWNT (Figure 6.7a) shows that most of the sidewalls are unfunctionalized, while Au-thiophene-SWNT (Figure 6.7c) shows long bands of continuous functionalized regions on SWNT. This could lead to the estimation that thiophene is better functionalized to SWNT than thiol. Yet, if we look up to the STM image (Figure 6.7b and d), in thiol-SWNTs the multiple functional groups are tightly bonded in about 5 - 25 nm, while in thiophene-SWNTs the functionalization is spread out uniformly along the whole length of SWNT. This information indicates that actually the functionalization levels of thiol- and thiophene-SWNTs are comparable. The difference is that, in thiol-SWNTs, functional groups are grouped together, and each group is bonded to a single gold nanoparticle, while in thiophene-SWNTs, every individual functional group is bonded to a nanoparticle.

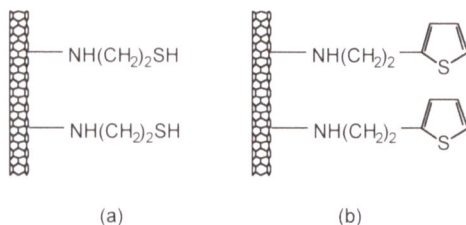

(a)   (b)

Figure 6.6: Structure of (a) thiol-functionalized SWNTs and thiophene-functionalized SWNTs.

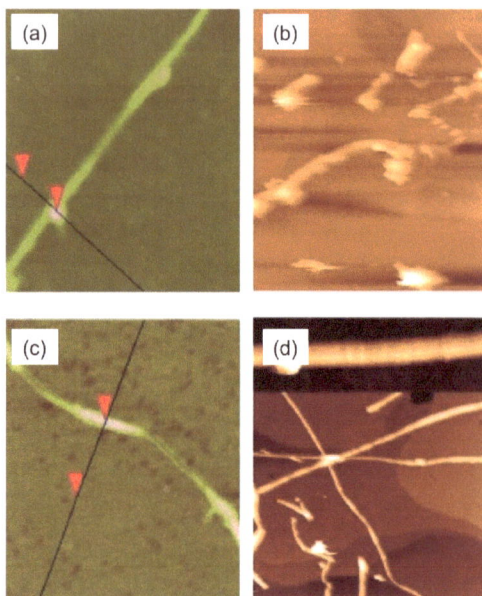

Figure 6.7: Difference between tapping mode AFM (a and c) and constant current mode STM (b and d, 4560 x 4000 Å) images of thiol- and thiophene-SWNTs (Figure 6.6). Inset in (d) is a higher resolution image of the local defects on thiophene-SWNT (500 x 140 Å). Adapted from L. Zhang, J. Zhang, N. Schmandt, J. Cratty, V. N. Khabashesku, K. F. Kelly, and A. R. Barron, AFM and STM characterization of thiol and thiophene functionalized SWNTs: pitfalls in the use of chemical markers to determine the extent of sidewall functionalization in SWNTs. *Chem. Commun.*, 2005, 5429. Copyright: Royal Society of Chemistry (2005).

Spin-polarized STM

Spin-polarized scanning tunneling microscopy (SP-STM) can be used to provide detailed information of magnetic phenomena on the single-atom scale. This imaging technique is particularly important for accurate measurement of superconductivity and high-density magnetic data storage devices. In addition, SP-STM, while sensitive to the partial magnetic moments of the sample, is not a field-sensitive technique and so can be applied in a variety of different magnetic fields.

Device setup and sample preparation

In SP-STM, the STM tip is coated with a thin layer of magnetic material. As with STM, voltage is then applied between tip and sample resulting in tunneling current. Atoms with partial magnetic moments that are aligned in the same direction as the partial magnetic moment of the atom at the very tip of the STM tip show a higher magnitude of tunneling current due to the interactions between the magnetic moments. Likewise, atoms with partial magnetic moments opposite that of the atom at the tip of the STM tip demonstrate a reduced tunneling current (Figure 6.8). A computer program can then translate the change in tunneling current to a topographical map, showing the spin density on the surface of the sample.

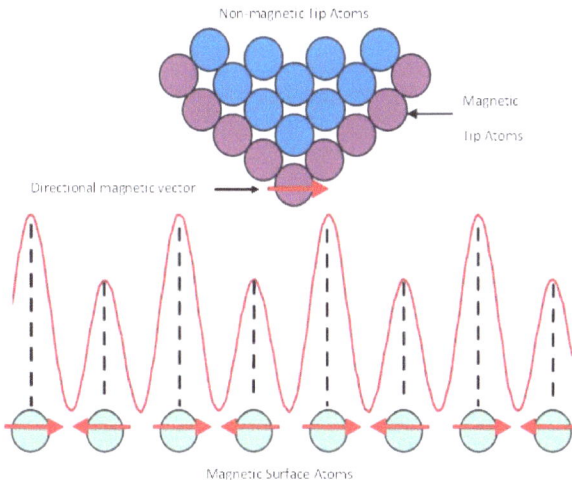

Figure 6.8: Schematic illustration of magnetized tip for SP-STM.

The sensitivity to magnetic moments depends greatly upon the direction of the magnetic moment of the tip, which can be controlled by the magnetic properties of the material used to coat the outermost layer of the tungsten STM probe. A wide variety of magnetic materials have been studied as possible coatings, including both ferromagnetic materials, such as a thin coat of iron or of gadolinium, and antiferromagnetic materials such as chromium. Another method that has been used to make a magnetically sensitive probe tip is irradiation of a semiconducting GaAs tip with high energy circularly polarized light. This irradiation causes a splitting of electrons in the GaAs valence band and population of the conduction band with spin-polarized electrons. These spin-polarized electrons then provide partial magnetic moments which in turn influence the tunneling current generated by the sample surface.

Sample preparation for SP-STM is essentially the same as for STM. SP-STM has been used to image samples such as thin films and nanoparticle constructs as well as determining the magnetic topography of thin metallic sheets such as in Figure 6.9. The upper image is a traditional STM image of a thin layer of cobalt, which shows the topography of the sample. The second image is an SP-STM image of the same layer of cobalt, which shows the magnetic domain of the sample. The two images, when combined provide useful information about the exact location of the partial magnetic moments within the sample.

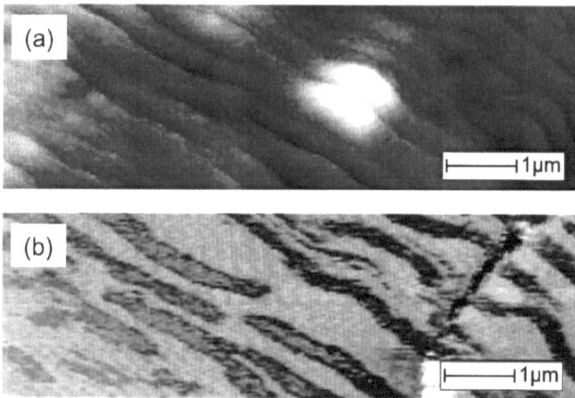

Figure 6.9: A thin layer of Co(0001) as imaged by (a) STM, showing the topography, and (b) SP-STM, showing the magnetic domain structure. Adapted from W. Wulfhekel and J. Kirschner, Spin-polarized scanning tunneling microscopy on ferromagnets. *Appl. Phys. Lett.*, 1999, 75, 1944. Copyright: American Institute of Physics (1999).

Limitations

One of the major limitations with SP-STM is that both distance and partial magnetic moment yield the same contrast in a SP-STM image. This can be corrected by combination with conventional STM to get multi-domain structures and/or topological information which can then be overlaid on top of the SP-STM image, correcting for differences in sample height as opposed to magnetization.

The properties of the magnetic tip dictate much of the properties of the technique itself. If the outermost atom of the tip is not properly magnetized, the technique will yield no more information than a traditional STM. The direction of the magnetization vector of the tip is also of great importance. If the magnetization vector of the tip is perpendicular to the magnetization vector of the sample, there will be no spin contrast. It is therefore important to carefully choose the coating applied to the tungsten STM tip in order to align appropriately with the expected magnetic moments of the sample. Also, the coating makes the magnetic tips more expensive to produce than standard STM tips. In addition, these tips are often made of mechanically soft materials, causing them to wear quickly and require a high cost of maintenance.

Ballistic electron emission microscopy

Ballistic electron emission microscopy (BEEM) is a technique commonly used to image semiconductor interfaces. Conventional surface probe techniques can provide detailed information on the formation of interfaces but lack the ability to study fully formed interfaces due to inaccessibility to the surface. BEEM allows for the ability to obtain a quantitative measure of electron transport across fully formed interfaces, something necessary for many industrial applications.

Device setup and sample preparation

BEEM utilizes STM with a three-electrode configuration, as seen in Figure 6.10. In this technique, ballistic electrons are first injected from a STM tip into the sample, traditionally composed of at least two layers separated by an interface, which rests on three indium contact pads that provide a connection to a base electrode (Figure 6.10). As the voltage is applied to the sample, electrons tunnel across the vacuum and through the first layer of the sample, reaching the interface, and then scatter. Depending on the magnitude of the voltage, some percentage of the electrons tunnel through the interface, and can be collected and measured as a current at a collector attached to the other side of the sample. The voltage from the STM tip is then varied, allowing for measurement of the barrier height. The barrier height is defined as the threshold at which electrons will cross the interface and are measurable as a current in the far collector. At a metal/n-type semiconductor interface this is the difference between the conduction band minimum and the Fermi level. At a metal/p-type semiconductor interface this is the difference between the valence band maximum of the semiconductor and the metal Fermi level. If the voltage is less than the barrier height, no electrons will cross the interface and the collector will read zero. If the voltage is greater than the barrier height, useful information can be gathered about the magnitude of the current at the collector as opposed to the initial voltage.

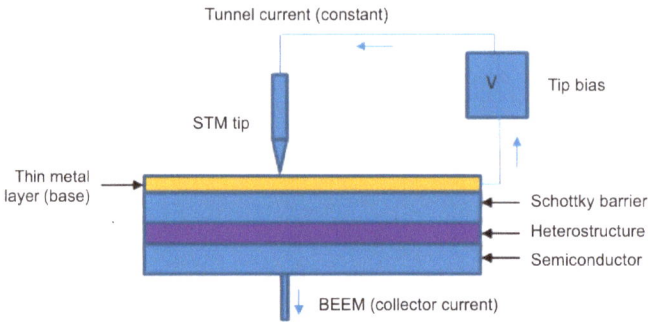

Figure 6.10: Diagram of a STM/BEEM system. The tip is maintained at the tunneling voltage, V, and the tunneling current, $I_t = V_I/R_F$, is held constant by the STM feedback circuit. The sample base layer is grounded and current into the semiconductor is measured by a virtual ground current amplifier.

Samples are prepared from semiconductor wafers by chemical oxide growth-strip cycles, ending with the growth of a protective oxide layer. Immediately prior to imaging the sample is spin-etched in an inert environment to remove oxides of oxides and then transferred directly to the ultra-high vacuum without air exposure. The BEEM apparatus itself is operated in a glove box under inert atmosphere and shielded from light.

Nearly any type of semiconductor interface can be imaged with BEEM. This includes both simple binary interfaces such as Au/n-Si(100) and more chemically complex interfaces such as Au/n-GaAs(100), such as seen in Figure 6.11.

(a)

(b)

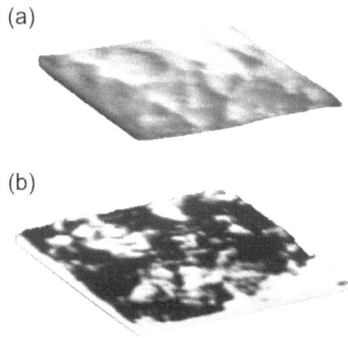

Figure 6.11: Images of Au/n-GaAs(100) layer (image area 510 Å x 390 Å) showing (a) the topography of the Au surface and (b) the BEEM grey-scale interface image. Image adapted from M. H. Hecht, L. D. Bell, W. J. Kaiser, and F. J. Grunthaner, Ballistic-electron-emission microscopy investigation of Schottky barrier interface formation. *Appl. Phys. Lett.*, **1989, 55, 780. Copyright: American Institute of Physics (1999).**

Limitations

Expected barrier height matters a great deal in the desired setup of the BEEM apparatus. If it is necessary to measure small collector currents, such as with an interface of high-barrier-height, a high-gain, low-noise current preamplifier can be added to the system. If the interface is of low-barrier-height, the BEEM apparatus can be operated at very low temperatures, accomplished by immersion of the STM tip in liquid nitrogen and enclosure of the BEEM apparatus in a nitrogen-purged glove box.

Photon STM

Photon scanning tunneling microscopy (PSTM) measures light to determine more information about characteristic sample topography. It has primarily been used as a technique to measure the electromagnetic interaction of two metallic objects in close proximity to one another and biological samples, which are both difficult to measure using many other common surface analysis techniques.

Device setup and sample preparation

This technique works by measuring the tunneling of photons to an optical tip. The source of these photons is the evanescent field generated by the total internal reflection (TIR) of a light beam from the surface of the sample (Figure 6.12). This field is characteristic of the sample material on the TIR surface, and can be measured by a sharpened optical fiber probe tip where the light intensity is converted to an electrical signal (Figure 6.13). Much like conventional STM, the force of this electrical signal modifies the location of the tip in relation to the sample. By mapping these modifications across the entire sample, the topography can be determined to a very accurate degree as well as allowing for calculations of polarization, emission direction and emission time.

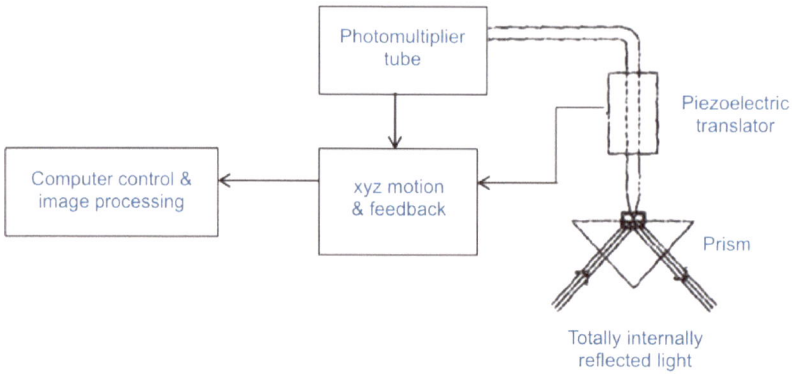

Figure 6.12: A schematic of a PSTM system.

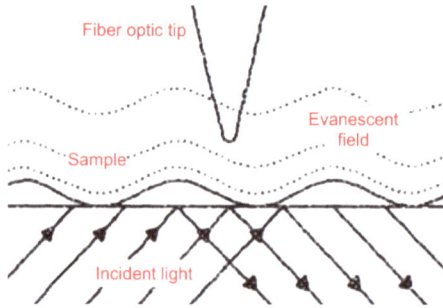

Figure 6.13: A TIR light beam generates an evanescent field which is modulated by the sample. A sharpened fiber optic probe tip receives light from the evanescent field and spatial variations in evanescent field intensity form the basis for imaging.

In PSTM, the vertical resolution is governed only by the noise, as opposed to conventional STM where the vertical resolution is limited by the tip dimensions. Therefore, this technique provides advantages over more conventional STM apparatus for samples where subwavelength resolution in the vertical dimension is a critical measurement, including fractal metal colloid clusters, nanostructured materials and simple organic molecules.

Samples are prepared by placement on a quartz or glass slide coupled to the TIR face of a triangular prism containing a laser beam, making the sample surface into the TIR surface (Figure 6.13). The optical fiber probe tips are constructed from UV grade quartz optical fibers by etching in HF acid to have nominal end diameters of 200 nm or less and resemble either a truncated cone or a paraboloid of revolution (Figure 6.14).

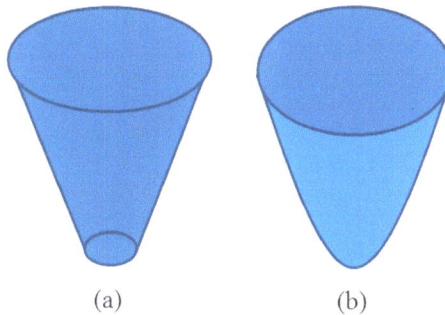

(a)   (b)

Figure 6.14: Possible optical fiber tip configurations: (a) truncated cone and (b) paraboloid of rotation.

PSTM shows much promise in the imaging of biological materials due to the increase in vertical resolution and the ability to measure a sample within a liquid environment with a high index TIR substrate and probe tip. This would provide much more detailed information about small organisms than is currently available.

Limitations

The majority of the limitations in this technique come from the materials and construction of the optical fibers and the prism used in the sample collection. The sample needs to be kept at low temperatures, typically around 100K, for the duration of the imaging and therefore cannot decompose or be otherwise negatively impacted by drastic temperature changes.

Scanning transmission electron microscope-electron energy loss spectroscopy

History of STEM-EELS

STEM-EELS is a terminology abbreviation for scanning transmission electron microscopy (STEM) coupled with electron energy loss spectroscopy (EELS). It works by combining two instruments, obtaining an image through STEM and applying EELS to detect signals on the specific selected area of the image. Therefore, it can be applied for a wide range of research such as characterizing morphology, detecting different elements, and different valence state. The first STEM was built by Baron Manfred von Arden (Figure 6.15) in around 1938, since it was just the prototype of STEM, it was not as good as transmission electron microscopy (TEM) by that time.

Figure 6.15: German physicist and inventor Baron Manfred von Arden (1907–1997).

Development of STEM was stagnant until the field emission gun was invented by Albert Crewe (Figure 6.16) in 1970s; he also came with the idea of annular dark field detector to detect atoms. In 1997, its resolution increased to 1.9 Å, and further increased to 1.36 Å in 2000. 4D STEM-EELS was developed recently, and this type of 4D STEM-EELS has high brightness STEM equipped with a high acquisition rate EELS detector, and a rotation holder. The rotation holder plays quite an important role to achieve this 4D aim, because it makes observation of the sample in 360° possible, the sample could be rotated to acquire the sample's thickness. High acquisition rate EELS enables this instrument the acquisition of the pixel spectrum in a few minutes.

Figure 6.16: British physicist Albert Crewe (1927 - 2009).

Basics of STEM-EELS

Interaction between electrons and sample

When electrons interact with the samples, the interaction between those two can be classified into two types, namely, elastic and inelastic interactions (Figure 6.17). In the elastic interaction, if electrons do not interact with the sample and pass through it, these electrons will contribute to the direct beam. The direct beam can be applied in STEM. In another case, electrons' moving direction in the sample is guided by the Coulombic force; the strength of the force is decided by charge and the distance between electrons and the core. In both cases, these is no energy transfer from electrons to the samples, that's the reason why it is called elastic interaction. In inelastic interaction, energy transfers from incident electrons to the samples, thereby, losing energy. The lost energy can be measured and how many electrons amounted to this energy can also be measured, and these data yield the electron energy loss spectrum (EELS).

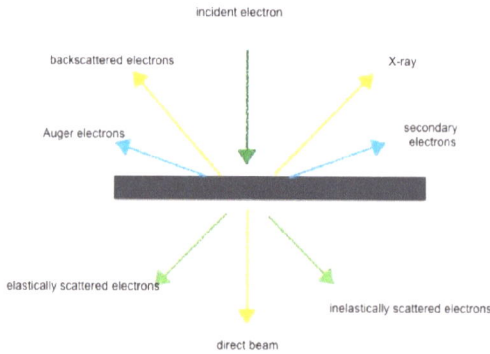

Figure 6.17: Demonstration of interaction between sample and electrons.

How do TEM, STEM and STEM-EELS work?

In transmission electron microscopy (TEM), a beam of electrons is emitted from tungsten source and then accelerated by electromagnetic field. Then with the aid of lens condenser, the beam will focus on and pass through the sample. Finally, the electrons will be detected by a charge-coupled device (CCD) and produce images, Figure 6.18.

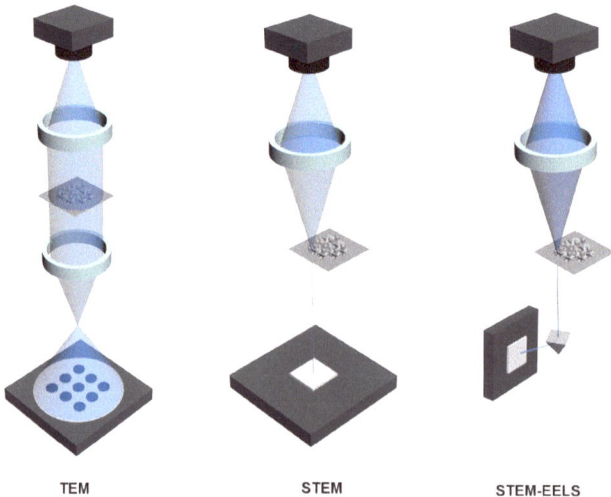

Figure 6.18: Scheme of TEM, STEM and STEM-EELS experiments. Adapted from http://toutestquantique.fr/en/scanning-electron/.

STEM works differently from TEM, the electron beam focuses on a specific spot of the sample and then raster scans the sample pixel by pixel, the detector will collect the transmitted electrons and visualize the sample. Moreover, STEM-EELS allows to analyze these electrons, the transmitted electrons could be characterized by adding a magnetic prism, the more energy the electrons lose, the more they will be deflected. Therefore, STEM-EELS can be used to characterize the chemical properties of thin samples.

Principle of STEM-EELS

A brief illustration of STEM-EELS is displayed in Figure 6.19. The electron source provides electrons, and it usually comes from a tungsten source located in a strong electrical field. The electron field will provide electrons with high energy. The condenser and the object lens also promote electrons forming into a fine probe and then raster scanning the specimen. The diameter of the probe will influence STEM's spatial resolution, which is caused by the lens aberrations. Lens aberration results from the refraction difference between light rays striking the edge and center point of the lens, and it also can happen when the light rays pass through with different energy. Base on this, an aberration corrector is applied to increase the objective aperture, and the incident probe will converge and increase the resolution, then promote sensitivity to single atoms. For the annular electron detector, the installment sequence of detectors is a bright field detector, a dark field detector and a high angle annular dark field detector. Bright field detector detects the direct beam that transmits through the specimen. Annular dark field detector collects the scattered electrons, which only go through at an aperture. The advantage of this is that it will not influence the EELS to detect signals from direct beam. High angle annular dark field detector collects electrons which are Rutherford scattering (elastic scattering of charged electrons), and its signal intensity is related with the square of atomic number (Z). So, it is also named as Z-contrast image. The unique point about STEM in acquiring image is that the pixels in image are obtained in a point by point mode by scanning the probe. EELS analysis is based on the energy loss of the transmitted electrons, so the thickness of the specimen will influence the detecting signal. In other words, if the specimen is too thick, the intensity of plasmon signal will decrease and may cause difficulty distinguishing these signals from the background.

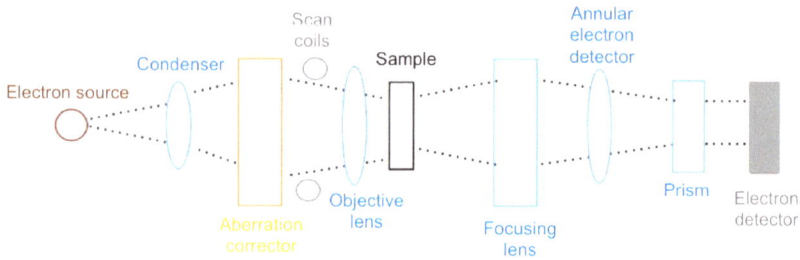

Figure 6.19: Schematic representation of STEM-EELS.

Typical features of EELS spectra

As shown in Figure 6.20, a significant peak appears at energy zero in EELS spectra and is therefore called zero-loss peak. Zero-loss peak represents the electrons which undergo elastic scattering during the interaction with specimen. Zero-loss peak can be used to determine the thickness of specimen according to,

$$t = \lambda_{inel} \ln[I_t/I_{ZLP}]$$

where t stands for the thickness, λinel is inelastic mean free path, It stands for the total intensity of the spectrum and IZLP is the intensity of zero loss peak.

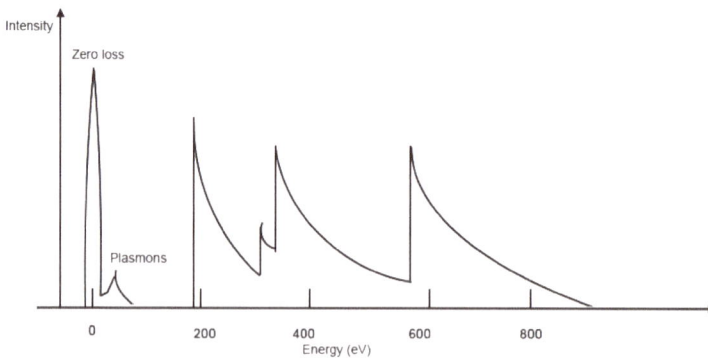

Figure 6.20: Typical features of EELS spectra. Adapted from http://www.mar-dre.com/homepage/mic/tem/eels/sld001.html.

The low loss region is also called valence EELS. In this region, valence electrons will be excited to the conduction band. Valence EELS can provide the information about band structure, bandgap, and optical properties. In the low loss region, plasmon peak is the most important. Plasmon is a phenomenon originates from the collective oscillation of weakly bound electrons. Thickness of the sample will influence the plasmon peak. The incident electrons will go through inelastic scattering several times when they interact with a very thick sample, and then result in convoluted plasmon peaks. It is also the reason why STEM-EELS favors sample with low thickness (usually less than 100 nm).

The high loss region is characterized by the rapidly increasing intensity with a gradually falling, which called ionization edge. The onset of ionization edges equals to the energy that inner shell electron needs to be excited from the ground state to the lowest unoccupied state. The amount of energy is unique for different shells and elements. Thus, this information will help to understand the bonding, valence state, composition and coordination information.

Energy resolution affects the signal to background ratio in the low loss region and is used to evaluate EELS spectrum. Energy resolution is based on the full width at half maximum of zero-loss peak.

Background signal in the core-loss region is caused by plasmon peaks and core-loss edges, and can be described by the following power law,

$$I_{BG} = AE^{-r}$$

where IBG stands for the background signal, E is the energy loss, A is the scaling constant and r is the slope exponent.

Therefore, when quantification the spectra data, the background signal can be removed by fitting pre-edge region with the above-mentioned equation and extrapolating it to the post-edge region.

Advantages and disadvantages of STEM-EELS

STEM-EELS has advantages over other instruments, such as the acquisition of high resolution of images. For example, the operation of TEM on samples sometimes result in blurring image and low contrast because of chromatic aberration. STEM-EELS equipped with aberration corrector, will help to reduce the chromatic aberration and obtain high quality image even at atomic resolution. It is very direct and convenient to understand the electron distributions on surface and bonding information. STEM-EELS also has the advantages in controlling the spread of energy. So, it becomes much easier to study the ionization edge of different material.

Even though STEM-EELS does bring a lot of convenience for research in atomic level, it still has limitations to overcome. One of the main limitations of STEM-EELS is controlling the thickness of the sample. As discussed above, EELS detects the energy loss of electrons when they interact with samples and the specimen, then the thickness of samples will impact on the energy lost detection. Simplify, if the sample is too thick, then most of the electrons will interact with the sample, signal to background ratio and edge visibility will decrease. Thus, it will be hard to tell the chemical state of the element. Another limitation is due to EELS needs to characterize low-loss energy electrons, which high vacuum condition is essential for characterization. To achieve such a high vacuum environment, high voltage is necessary. STEM-EELS also requires the sample substrates to be conductive and flat.

Application of STEM-EELS

STEM-EELS can be used to detect the size and distribution of nanoparticles on a surface. For example, CoO on MgO catalyst nanoparticles may be prepared by hydrothermal methods. The size and distribution of nanoparticles will greatly influence the catalytic properties, and the distribution and morphology change of CoO nanoparticles on MgO is important to understand. Co L_3/L_2 ratios display uniformly around 2.9 (Figure 6.21), suggesting that Co^{2+} dominates the electron state of Co. The results show that the ratios of O:(Co+Mg) and Mg:(Co+Mg) are not consistence, indicating that these three elements are in a random distribution. STEM-EELS mapping images results further confirm the non-uniformity of the elemental distribution, consistent with a random distribution of CoO on the MgO surface (Figure 6.22).

Figure 6.23 shows the K-edge absorption of carbon and transition state information could be concluded. Typical carbon-based materials have the features of the transition state, such that 1s transits to π^* state and 1s to σ^* states locate at 285 and 292 eV, respectively. The two-transition state correspond to the electrons in the valence band electrons being excited to conduction state. Epoxy exhibits a sharp peak around 285.3 eV compared to GO and GNPs. Meanwhile, GNPs have the sharpest peak around 292 eV, suggesting the most

C atoms in GNPs are in 1s to σ* state. Even though GO is in oxidation state, part of its carbon still behaves 1s transits to π*.

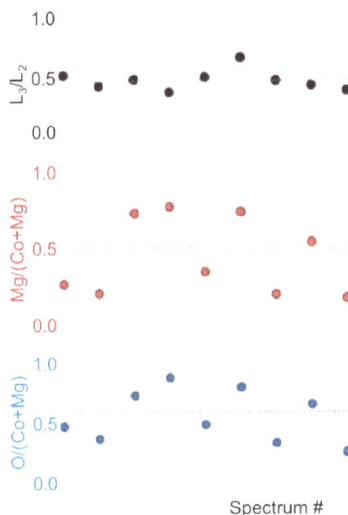

Figure 6.21: EELS signal ratio of Co L₃/L₂, and O and Mg EELS signals relative to combined Co + Mg signals for a CoO/MgO sample. Adapted from S. Alayoglu, D. J. Rosenberg, and M. Ahmed, Hydrothermal synthesis and characterization under dynamic conditions of cobalt oxide nanoparticles supported over magnesium oxide nano-plates. *Dalton Trans.*, 2016, 45, 9932. Copyright: Royal Society of Chemistry (2016).

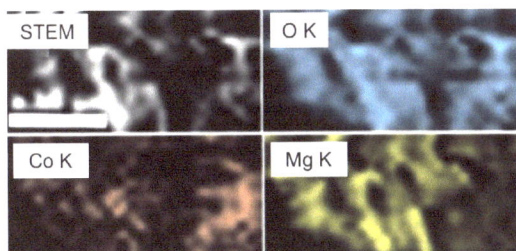

Figure 6.22: STEM image and EELS maps acquired at O K, Co L and Mg K edges for a CoO/MgO sample. Adapted from S. Alayoglu, D. J. Rosenberg, and M. Ahmed, Hydrothermal synthesis and characterization under dynamic conditions of cobalt oxide nanoparticles supported over magnesium oxide nano-plates. *Dalton Trans.*, 2016, 45, 9932. Copyright: Royal Society of Chemistry (2016).

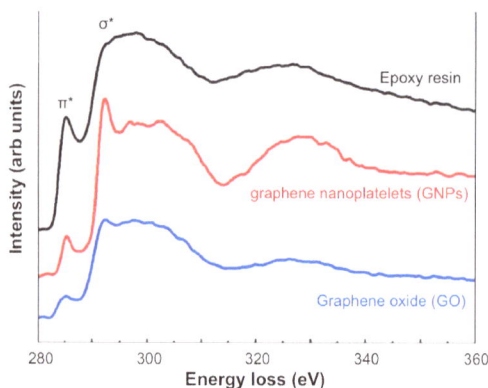

Figure 6.23: EELS spectrum of graphene nanoplatelets (GNPs), graphene oxide (GO) in comparison with an epoxide resin. Adapted from Y. Liu, A. L. Hamon, P. Haghi-Ashtiani, T. Reiss, B. Fan, D. He, and J. Bai, Quantitative study of interface/interphase in epoxy/graphene-basednanocomposites by combining STEM and EELS. *ACS Appl. Mater. Inter.*, 2016, 8, 34151. Copyright: American Chemical Society (2017).

The annular dark filed (ADF) mode of STEM provides information about atomic number of the elements in a sample. For example, the ADF image (Figure 6.24) of $La_{1.2}Sr_{1.8}Mn_2O_7$ (Figure 6.25) along [010] direction shows bright spots and dark spots, and even for bright spots (p and r), they display different levels of brightness.

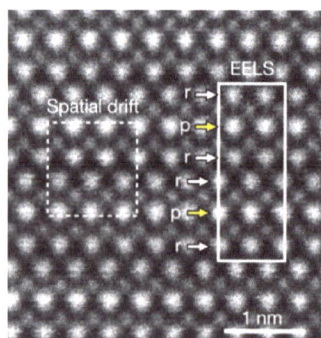

Figure 6.24: ADF image of of $La_{1.2}Sr_{21.8}Mn_2O_7$, observed along the [010] direction. Adapted from K. Kimoto, T. Asaka, T. Nagai, M. Saito, Y. Matsui, K. Ishizuka, Element-selective imaging of atomic columns in a crystal using STEM and EELS. *Nature*, 2007, 450, 702. Copyright: Nature Publishing Group (2007).

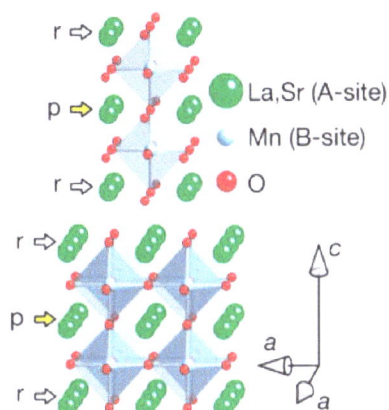

Figure 6.25: Crystal structure of La$_{1.2}$Sr$_{21.8}$Mn$_2$O$_7$. Adapted from K. Kimoto, T. Asaka, T. Nagai, M. Saito, Y. Matsui, K. Ishizuka, Element-selective imaging of atomic columns in a crystal using STEM and EELS. *Nature*, 2007, 450, 702. Copyright: Nature Publishing Group (2007).

This phenomenon is caused by the difference in atomic numbers. Bright spots are La and Sr, respectively. Dark spots are Mn elements. O is too light to show on the image. EELS result shows the core-loss edge of La, Mn and O (Figure 6.26), but the researchers did not give information on core-loss edge of Sr, Sr has N$_{2,3}$ edge at 29 eV and L$_3$ edge at 1930 eV and L$_2$ edge at 2010 eV.

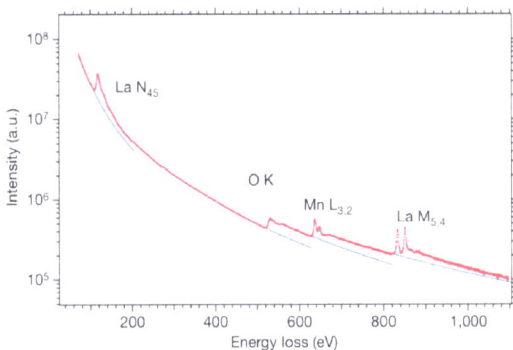

Figure 6.26: STEM-EELS data of La$_{1.2}$Sr$_{21.8}$Mn$_2$O$_7$ obtained from rectangular area of (Figure 6.24) and the blue area equals the core-loss of each element. Adapted from K. Kimoto, T. Asaka, T. Nagai, M. Saito, Y. Matsui, K. Ishizuka, Element-selective imaging of atomic columns in a crystal using STEM and EELS. *Nature*, 2007, 450, 702. Copyright: Nature Publishing Group (2007).

Bibliography

S. Alayoglu, D. J. Rosenberg, and M. Ahmed, Hydrothermal synthesis and character-ization under dynamic conditions of cobalt oxide nanoparticles supported over magnesium oxide nano-plates. *Dalton Trans.*, 2016, **45**, 9932.

R. Berndt and J. K. Gimzewski, Photon emission in scanning tunneling microscopy: Interpretation of photon maps of metallic systems. *Phys. Rev. B*, 1993, **48**, 4746.

G. Binnig and H. Rohrer, Scanning tunneling microscopy. *Surf. Sci.*, 1983, **126**, 236.

G. Binnig, H. Rohrer, C. Gerber, and E. Weibel, Surface studies by scanning tunnel-ing microscopy. *Phys. Rev. Lett.*, 1982, **49**, 57.

M. Bode, Spin-polarized scanning tunnelling microscopy. *Rep. Prog. Phys.* 2003, **66**, 523.

R. F. Egerton, Electron energy-loss spectroscopy in the TEM. *Rep Prog Phys,* 2008, **72**, 016502.

J. Griffith, Scanning tunneling microscopy. *Annu. Rev. Mater. Sci.*, 1990, **20**, 219.

M. H. Hecht, L. D. Bell, W. J. Kaiser, and F. J. Grunthaner, Ballistic-electron-emis-sion microscopy investigation of Schottky barrier interface formation. *Appl. Phys. Lett.*, 1989, **55**, 780.

K. Jarausch, P. Thomas, D.N. Leonard, R. Twesten and C. R. Booth, Four-dimen-sional STEM-EELS: Enabling nano-scale chemical tomography. *Ultramicroscopy,* 2009, **109**, 326.

W. J. Kaiser and L. D. Bell, Direct investigation of subsurface interface electronic structure by ballistic-electron-emission microscopy. *Phys. Rev. Lett.*, 1988, **60**, 1406.

K. Kimoto, T. Asaka, T. Nagai, M. Saito, Y. Matsui, K. Ishizuka, Element-selective imaging of atomic columns in a crystal using STEM and EELS. *Nature*, 2007, **450**, 702

Y. Liu, A. L. Hamon, P. Haghi-Ashtiani, T. Reiss, B. Fan, D. He, and J. Bai, Quanti-tative study of interface/interphase in epoxy/graphene-basednanocomposites by combining STEM and EELS. *ACS Appl. Mater. Inter.*, 2016, **8**, 34151

S. Marchini, S. Gunther, and J. Wintterlin, Scanning tunneling microscopy of gra-phene on Ru(0001). *Phys. Rev. B*, 2007, **76**, 075429.

M. Poggi, L. Bottomley, and P. Lillehei, Scanning probe microscopy. *Anal. Chem.*, 2002, **74**, 2851.

R. C. Reddick, D. W. Warmack, D. W. Chilcott, S. L. Sharp, and T. L. Ferrell, Photon scanning tunneling microscopy. *Rev. Sci. Instrum.*, 1990, **61**, 3669.

P. Samori, Exploring supramolecular interactions and architectures by scanning force microscopies. *J. Mater. Chem.*, 2005, **14**, 1353.

D. P. Tsai, J. Kovacs, Z. Wang, M. Moskovits, V. M. Shalaev, J. S. Suh, and R. Botet, Photon scanning tunneling microscopy images of optical excitations of fractal metal colloid clusters. *Phys. Rev. Lett.*, 1994, **72**, 4149.

W. Wulfhekel and J. Kirschner, Spin-polarized scanning tunneling microscopy on ferromagnets. *Appl. Phys. Lett.*, 1999, **75**, 1944.

L. Zhang, J. Zhang, N. Schmandt, J. Cratty, V. N. Khabashesku, K. F. Kelly, and A. R. Barron, AFM and STM characterization of thiol and thiophene functionalized SWNTs: pitfalls in the use of chemical markers to determine the extent of side-wall functionalization in SWNTs. *Chem. Commun.*, 2005, 5429.

Chapter 7: Atomic Force Microscopy

Samuel Maguire-Boyle, Hannah Rutledge,
Brittany L. Oliva-Chatelain, Ryan Thaner, Lin Yuan,
Pavan M. V. Raja and Andrew R. Barron

Introduction

Atomic force microscopy (AFM) is a high-resolution form of scanning probe microscopy, also known as scanning force microscopy (SFM). The instrument uses a cantilever with a sharp tip at the end to scan over the sample surface (Figure 7.1). As the probe scans over the sample surface, attractive or repulsive forces between the tip and sample, usually in the form of van der Waal forces but also can be a number of others such as electrostatic and hydrophobic/hydrophilic, cause a deflection of the cantilever. The deflection is measured by a laser (Figure 7.1) which is reflected off the cantilever into photodiodes. As one of the photodiodes collects more light, it creates an output signal that is processed and provides information about the vertical bending of the cantilever. This data is then sent to a scanner that controls the height of the probe as it moves across the surface. The variance in height applied by the scanner can then be used to produce a three-dimensional topographical representation of the sample.

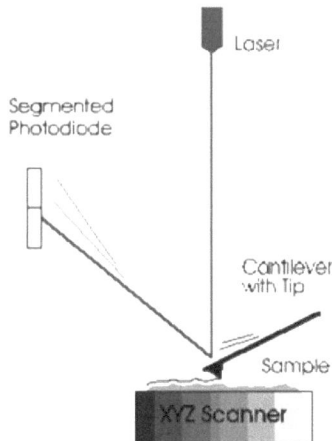

Figure 7.1: Simple schematic of atomic force microscope (AFM) apparatus. Adapted from H. G. Hansma, Department of Physics, University of California, Santa Barbara.

Modes of operation

Contact mode

The contact mode method utilizes a constant force for tip-sample interactions by maintaining a constant tip deflection (Figure 7.2). The tip communicates the nature of the interactions that the probe is having at the surface via feedback loops and the scanner moves the entire probe in order to maintain the original deflection of the cantilever. The constant force is calculated and maintained by using Hooke's Law,

$$F = -kx$$

This equation relates the force (F), spring constant (k), and cantilever deflection (x). Force constants typically range from 0.01 to 1.0 N/m. Contact mode usually has the fastest scanning times but can deform the sample surface. It is also only the only mode that can attain "atomic resolution."

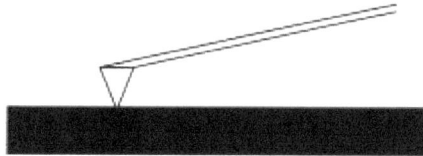

Figure 7.2: Schematic diagram of probe and surface interaction in contact mode.

Tapping mode

In the tapping mode the cantilever is externally oscillated at its fundamental resonance frequency (Figure 7.3). A piezoelectric on top of the cantilever is used to adjust the amplitude of oscillation as the probe scans across the surface. The deviations in the oscillation frequency or amplitude due to interactions between the probe and surface are measured and provide information about the surface or types of material present in the sample. This method is gentler than contact AFM since the tip is not dragged across the surface, but it does require longer scanning times. It also tends to provide higher lateral resolution than contact AFM.

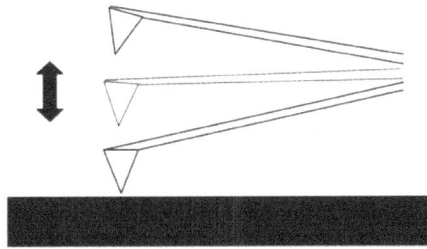

Figure 7.3: Diagram of probe and surface interaction in tapping mode.

Noncontact mode

For noncontact mode the cantilever is oscillated just above its resonance frequency and this frequency is decreased as the tip approaches the surface and experiences the forces associated with the material (Figure 7.4). The average tip-to-sample distance is measured as the oscillation frequency or amplitude is kept constant, which then can be used to image the surface. This method exerts very little force on the sample, which extends the lifetime of the tip. However, it usually does not provide very good resolution unless placed under a strong vacuum.

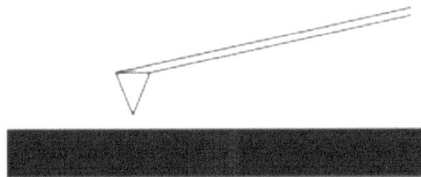

Figure 7.4: Diagram of probe and surface interaction in noncontact mode.

Experimental limitations

A common problem seen in AFM images is the presence of artifacts which are distortions of the actual topography, usually either due to issues with the probe, scanner, or image processing. The AFM scans slowly which makes it more susceptible to external temperature fluctuations leading to thermal drift. This leads to artifacts and inaccurate distances between topographical features.

It is also important to consider that the tip is not perfectly sharp and therefore may not provide the best aspect ratio, which leads to a convolution of the true topography. This leads to features appearing too large or too small since the width of the probe cannot precisely move around the particles and holes on the surface. It is for this reason that tips with smaller radii of curvature provide better resolution in imaging. The tip can also produce false images and poorly contrasted images if it is blunt or broken.

The movement of particles on the surface due to the movement of the canti-lever can cause noise, which forms streaks or bands in the image. Artifacts can also be made by the tip being of inadequate proportions compared to the surface being scanned. It is for this reason that it is important to use the ideal probe for the particular application.

Sample size and preparation

The sample size varies with the instrument, but a typical size is 8 mm by 8 mm with a typical height of 1 mm. Solid samples present a problem for AFM since the tip can shift the material as it scans the surface. Solutions or disper-sions are best for applying as uniform of a layer of material as possible in order to get the most accurate value of particles' heights. This is usually done by spin-coating the solution onto freshly cleaved mica which allows the par-ticles to stick to the surface once it has dried.

Applications of AFM

AFM is particularly versatile in its applications since it can be used in ambient temperatures and many different environments. It can be used in many differ-ent areas to analyze different kinds of samples such as semiconductors, polymers, nanoparticles, biotechnology, and cells amongst others. The most common application of AFM is for morphological studies in order to attain an understanding of the topography of the sample. Since it is common for the material to be in solution, AFM can also give the user an idea of the ability of the material to be dispersed as well as the homogeneity of the particles within that dispersion. It also can provide a lot of information about the particles being studied such as particle size, surface area, electrical properties, and chemical composition. Certain tips are capable of determining the principal mechanical, magnetic, and electrical properties of the material. For example, in magnetic force microscopy (MFM) the probe has a magnetic coating that senses magnetic, electrostatic, and atomic interactions with the surface. This type of scanning can be performed in static or dynamic mode and depicts the magnetic structure of the surface.

AFM of carbon nanotubes

Atomic force microscopy is usually used to study the topographical morphology of these materials. By measuring the thickness of the material, it is possible to determine if bundling occurred and to what degree. Other dimensions of the sample can also be measured such as the length and width of the tubes or bundles. It is also possible to detect impurities, functional groups (Figure 7.5), or remaining catalyst by studying the images. Various methods of producing nanotubes have been found and each demonstrates a slightly different profile of homogeneity and purity. These impurities can be carbon coated metal, amorphous carbon, or other allotropes of carbon such as fullerenes and graphite. These facts can be utilized to compare the purity and homogeneity of the samples made from different processes, as well as monitor these characteristics as different steps or reactions are performed on the material. The distance between the tip and the surface has proven itself to be an important parameter in noncontact mode AFM and has shown that if the tip is moved past the threshold distance, approximately 30 μm, it can move or damage the nanotubes. If this occurs, a useful characterization cannot be performed due to these distortions of the image.

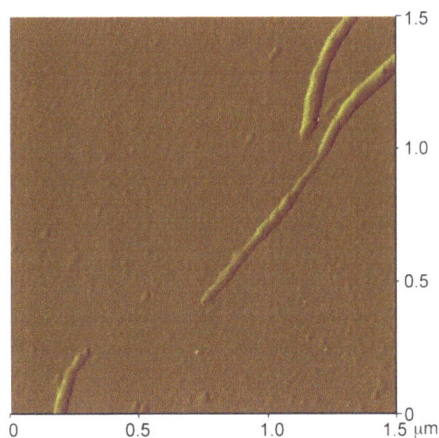

Figure 7.5: AFM image of a polyethyleneimine-functionalized single walled carbon nanotube (PEI-SWNT) showing the presence of PEI "globules" on the SWNT. Adapted from E. P. Dillon, C. A. Crouse, and A. R. Barron, Synthesis, characterization, and carbon dioxide adsorption of covalently attached polyethyleneimine-functionalized single-wall carbon nanotubes. *ACS Nano*, 2008, 2, 156. Copyright: American Chemical Society (2008).

AFM of fullerenes

Atomic force microscopy is best applied to aggregates of fullerenes rather than individual ones. While the AFM can accurately perform height analysis of individual fullerene molecules, it has poor lateral resolution and it is difficult to accurately depict the width of an individual molecule. Another common issue that arises with contact AFM and fullerene deposited films is that the tip shifts clusters of fullerenes which can lead to discontinuities in sample images.

Friction measurements

Atomic force microscopy also has the ability to measure frictional properties. The direct signal acquired is the current change caused due to the lateral force on the sample interacting with the tip, so the unit is usually nA. The topography and height profile are acquired using the same method in the tapping mode. However, there are two additional pieces of information that are necessary in order to determine the frictional properties of the material. First, the normal load. The normal load is described in,

$$F = C \times \Delta z$$

however, what we directly get here proportional to the normal load is the setpoint we give it for the tip to the sample. It is a current. So, we need a vertical force coefficient (C_{VF}) to get what the normal load we apply to the material, as illustrated in,

$$F = I_{setpoint} \times C_{VF}$$

As discuss above, the coefficient depends on the tip itself. It is shown in,

$$C_{VF} = K/L$$

where K is the stiffness of the tip, it can be got through the vibrational model of the cantilever, and usually we can get it if we buy the commercial AFM tip. L is the optical coefficient of the cantilever, it can be acquired by calibrate the force-displacement curve of the tip, as shown in Figure 7.6. Then L can be acquired by getting the slope of process 1 or 6 in Figure 7.6.

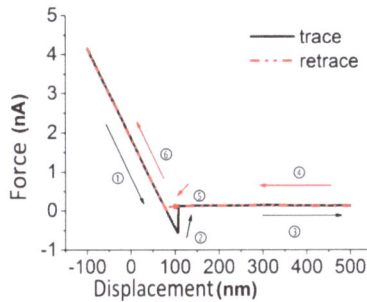

Figure 7.6: Force-displacement curve calibration of the tip.

Figure 7.7 is a typical friction image composed of n × n lines. Each point is the friction force value corresponding to that point. All we need to do is to get the average friction for the area we are interested in. Then use this current signal multiplied by the lateral force coefficient then we can obtain the actual friction force.

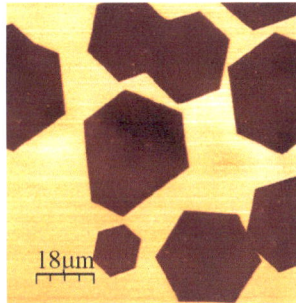

Figure 7.7: The friction image by AFM of the CVD grown monolayer graphene. Adapted from P. Egberts, G. H. Han, X. Z. Liu, A. T. C. Johnson, and R. W. Carpick, Frictional behavior of atomically thin sheets: hexagonal-shaped graphene islands grown on copper by chemical vapor deposition. *ACS Nano*, 2014, 8, 5010. Copyright: American Chemical Society (2014).

During the process of collecting the original data of the lateral force (friction), for every line in the image, the friction information is actually composed of two data line: trace and retrace (see Figure 7.6). The average of results for trace (Figure 7.6, black line) and retrace (Figure 7.6, red line) as the friction signal of the certain point on the line. That is to say, the actual friction is determined from,

$$F_f = \frac{I_{forward} - I_{backward}}{2} \times C_{LF}$$

where the $I_{forward}$ and $I_{backward}$ are data points we can derive from the trace and retrace from the friction image, and C_{LF} is the lateral force coefficient.

Data analysis

There are several ways to compare the details of the frictional properties at the nanoscale. Figure 7.8 is an example comparing the friction on the sample (in this case, few-layer graphene) and the friction on the substrate (SiO_2). Qualitatively we can easily see the friction on the graphene is way smaller than it on the SiO_2 substrate. As graphene is a great lubricant and have low friction, the original data just enable us to confirm that.

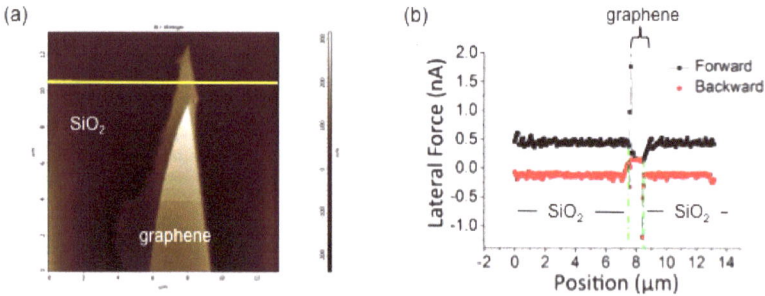

Figure 7.8: AFM image of few-layer graphene (a) and the friction profile (b) along the selected (yellow) line in (a).

Figure 7.9 shows multi-layers of graphene on a mica. By selecting a certain cross section line and comparing both height profile and friction profile, it will provide us some information of the friction related to the structure behind this section. The friction-distance curve is a typical important path for the data analysis.

We can also take the average of friction signal for an area and compare that from the region to the region. Figure 7.10 shows a region of the graphene with the layer numbers from 1-4. Figure 7.10a and b are also the topography and the friction image respectively. By compare the average friction from the area to the area, we can obviously see the friction on graphene decreases as the number of layers increases. Though Figure 7.10c and d we can obviously see this average friction change on the surface from 1 to 4 layers of graphene. But for a more general statistical way, getting the normalized signal of the average friction and comparing them can be more straightforward.

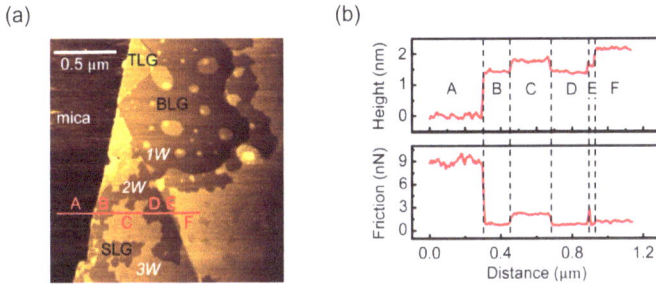

Figure 7.9: The topography of graphene on mica (a) and the corresponding height and friction profile (b) of the selected section defined by the red line in (a). Adapted from H. Lee, J. H. Ko, J. S. Choi, J. H. Hwang, Y. H. Kim, M. Salmeron, and J. Y. Park, Enhancement of friction by water intercalated between graphene and mica. *J. Phys. Chem. Lett.*, 2017, 8, 3482. Copyright: American Chemical Society (2017).

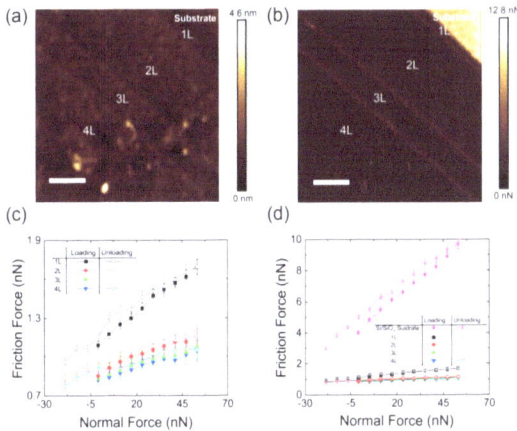

Figure 7.10: (a) The topography image of graphene from 1 to 4 layers on SiO$_x$. (b) The corresponding friction image of (a). (c) and (d) are the corresponding Friction-Normal Load curves of the area. Adapted from P. Gong, Z. Ye, L. Yuan, and P. Egberts, Evaluation of wetting transparency and surface energy of pristine and aged graphene through nanoscale friction. *Carbon*, 2018, 132, 749. Copyright: Elsevier (2018).

Another way to compare the frictional properties is that, to apply different normal load and see how the friction change, then get the information on friction-normal load curve. This is important because we know too much normal load for the materials can easily break or wear the materials. Examples and details will be discussed below.

The effect of H_2O: a cautionary tale

During the process of using tip approach to graphene and applying the normal load (increasing normal load, loading process) and withdrawing the tip gradually (decreasing normal load, unloading process), the friction on graphene exhibits hysteresis, which means a large increment of the friction while we drag off the tip. This process can be analyzed from friction-normal load curve, as shown in Figure 7.11. It was thought that this effect may be due to the detail of interacting behavior of the contact area between the tip and graphene. However, if you test this in different ambient conditions, for example if nitrogen was blown into the chamber while testing occured, this hysteresis disappears.

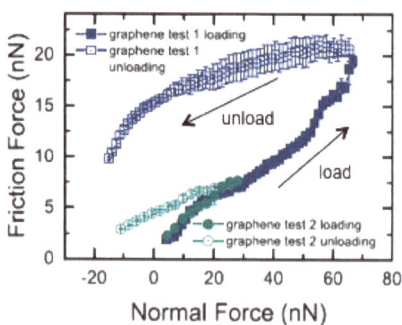

Figure 7.11: Friction hysteresis on the surface of graphene/Cu. Adapted from P. Egberts, G. H. Han, X. Z. Liu, A. T. C. Johnson, and R. W. Carpick, Frictional behavior of atomically thin sheets: hexagonal-shaped graphene islands grown on copper by chemical vapor deposition. *ACS Nano*, 2014, 8, 5010. Copyright: American Chemical Society (2014).

In order to explore the mechanism of such a phenomenon, a series of friction test under different conditions. A key factor here is the humidity in the testing environment. Figure 7.12 is a typical friction measurement on monolayer and 3-layer graphene on SiO_x. We can see the friction hysteresis is very different under dry nitrogen gas (0.1% humidity) and the ambient (24% humidity) from Figure 7.12.

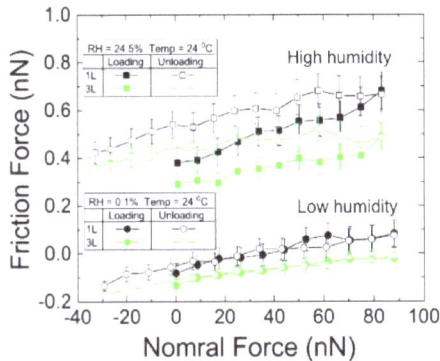

Figure 7.12: Friction behavior of monolayer and 3-layer graphene under different humidity conditions. P. Gong, Z. Ye, L. Yuan, and P. Egberts, Evaluation of wetting transparency and surface energy of pristine and aged graphene through nanoscale friction. *Carbon*, **2018, 132, 749. Copyright: Elsevier (2018).**

Simulation on this system suggests this friction hysteresis on the surface of graphene is due to the water interacting with the surface of graphene. The contact angle between the tip/water molecule-graphene interfaces is the key component. The further study suggests once you put the graphene samples in air and expose them for a long period of times (several days), the chemical bonding at the surface can change due to the water molecule in the air so that the friction properties at nanoscale can be very different.

The bond between the material under investigation and the substrate can be very vital for the friction behavior at the nanoscale. The studies during the years suggest that the friction of the graphene will decrease as the number of layers increase. This is adaptable for suspended graphene (with nothing to support it), and graphene on most of substrates (such as SiO_x, Cu foil and so on). However, if the graphene is supported by fresh cleaved mica surface, there's no difference for the frictional properties of different-layer graphene, this is due to the large surface dissipation energy, so the graphene is very firmly fixed to the mica.

However, on the other hand, the surface of mica is also hydrophilic, this is causal to the water distribution on the surface of mica, and the water intercalation between the graphene and mica bonding. Through the friction measurement of the graphene on mica, we can analyze this system quantitatively, as shown in Figure 7.12.

A practical guide to using the nanoscope atomic force microscope

The following is intended as a guide for use of the Nanoscope AFM system, however, it can be adapted for similar AFM instruments.

Initial setup

Turn on each component shown in Figure 7.13.

- The controller that powers the scope (the switch is at the back of the box).
- The camera monitor.
- The white light source.

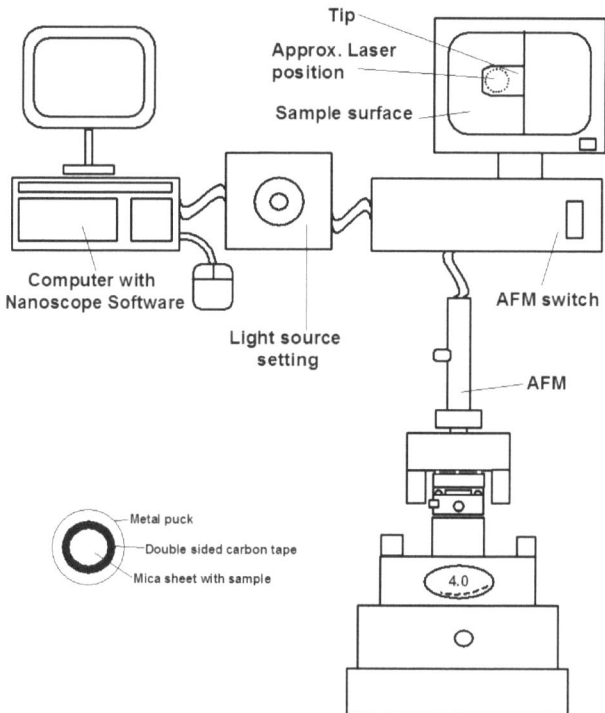

Figure 7.13: Schematic representation of the AFM computer, light source, camera set-up, and sample puck.

Select imaging mode using the mode selector switch is located on the left-hand side of the atomic force microscope (AFM) base (Figure 7.14), there are three modes:

- Scanning tunneling microscopy (STM).
- Atomic force microscopy/lateral force microscopy (AFM/LFM).
- Tapping mode atomic force microscopy (TM-AFM).

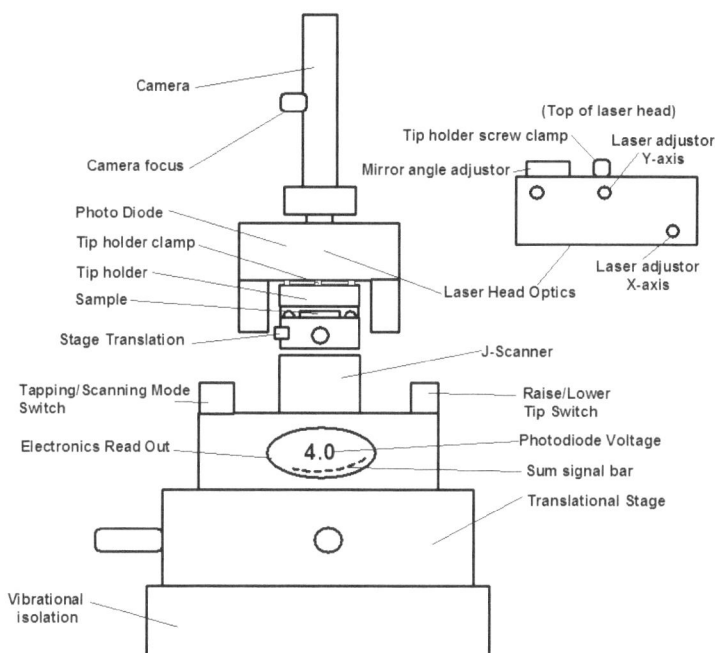

Figure 7.14: Schematic representation of the AFM.

Sample preparation

Most particulate samples are imaged by immobilizing them onto mica sheet, which is fixed to a metal puck. Samples that are in a solvent are easily deposited. To make a sample holder a sheet of mica is punched out and stuck to double-sided carbon tape on a metal puck. In order to ensure a pristine surface, the mica sheet is cleaved by removing the top sheet with Scotch™ tape to reveal a pristine layer underneath. The sample can be spin coated onto the mica or air dried.

The spin coat method

1. Use double-sided carbon sticky tape to secure the puck on the spin coater.
2. Load the sample by drop casting the sample solution onto the mica surface.
3. The sample must be dry to ensure that the tip remains clean.

Puck mounting

1. Place the sample puck in the magnetic sample holder and center the sample.
2. Verify that the AFM head is sufficiently raised to clear the sample with the probe. The sample plane is lower than the plane defined by the three balls. The sample should sit below the nubs. Use the lever on the right side of the J-scanner to adjust the height. (N.B. the labels up and down refer to the tip. "Tip up" moves the sample holder down to safety, and tip down moves the sample up. Use caution when moving the sample up.)
3. Select the appropriate cantilever for the desired imaging mode. The tips are fragile and expensive (*ca.* $20 per tip) so handle with care. Contact AFM use a silicon nitride tip (NP), while tapping AFM use a silicon tip (TESP).

Tip mounting and alignment

1. Mount a tip using the appropriate fine tweezers. Use the tweezers carefully to avoid possible misalignment. Work on a white surface (a piece of paper or a paper towel) so that the cantilever can be easily seen. The delicate part of the tip the cantilever is located at the beveled end and should not be handled at that end (shown in Figure 7.15). The tips are stored on a tacky gel tape. Use care, as dropping the tip will break the cantilever. Think carefully about how you approach the tip with the tweezers. Generally gripping it from the side is the best option. Once the tip is being held by the tweezers it needs to be placed in the tip holder clamp. With one hand holding the tweezers, use the other hand to open the clip by pressing down on the whole holder while it is lying on a flat hard surface. Once the clip is raised by downward pressure insert the tip (Figure 7.16a). Make sure the tip is seated firmly and that the back end is in contact with the end of the probe groove, there is a circular hole in the clamp. When the clamp holds the tip, the hole should look like a half moon, with half filled with the

back straight end of the tip. The groove is larger than the tip, so try to put the tip in the same place each time you replace it to improve reproducibility.

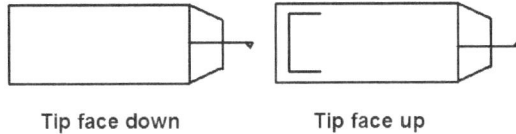

Tip face down   Tip face up

Figure 7.16: Schematic views of the AFM tip.

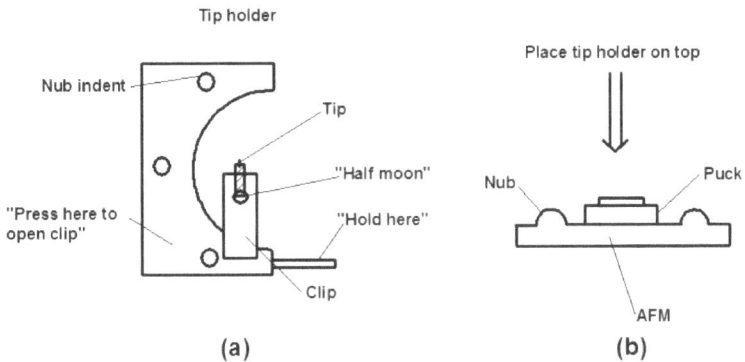

(a)   (b)

Figure 7.16: Schematic view of (a) the tip holder and (b) the tip holder location in the AFM.

2. Carefully place the tip holder onto the three nubs to gently hold it in place. Bring the holder in at angle to avoid scraping it against the sample (Figure 7.16b).
3. Tighten the clamping screw located on the back of the AFM head to secure the cantilever holder and to guarantee electrical contact. The screw is on the back of the laser head, at the center.
4. Find the cantilever on the video display. Move the translational stage to find it.
5. Adjust the focusing knob of the optical microscope (located above AFM head) to focus on the cantilever tip. Tightening the focus knob moves the camera up. Focus on the dark blob on the right-hand side of the screen as that is the cantilever.
6. Focus on the top mica surface, being careful not to focus on the bottom surface between the top of the double-sided carbon tape and the mica surface. Generally, you will see a bubble trapped between the

carbon tape and the mica surface. If you are focused on the top surface, you can frequently see the reflection of the tip on the mica
surface. The real focus is half-way between the two cantilever focus
points.

7. Slowly lower the tip down to the surface, if the camera is focused
properly onto the surface the cantilever tip will gradually come into
view. Keep lowering until the two tips images converge into one.
Please note that you can crash the tip into the surface if you go past
this point. This is damaging to the tip and may not be possible to obtain an image if it happens, and the tip may have to be replaced. You
will know if this happens when looking at the cantilever tip if it goes
from black to bright white. At this point the tip is in contact with the
surface and turns white as it is not reflecting light back into the photodiode but instead into the camera.

8. Find the laser spot, it the spot is not visible on the camera screen look
at the cantilever holder and see if it was visible. It helps to lower the
brightness of the white light, use the translational stage again to
search for it.

9. Once the laser spot has been located use the x and y laser adjustment
knobs to align the laser spot roughly onto the tip of the cantilever.

10. Maximize the sum signal using the photo-detector mirror lever located on the back of the head and the laser x and y translation. As long
as the sum signal value is above 3.6 V, the instrument will work, but
keep adjusting the X and Y directions of the laser until the sum signal
is as high as possible.

11. To ensure that the laser is centered on the photodiode, zero the detector signals using the mirror adjustment knobs located on the top and
back of the head. The knob on the top of the head adjusts TMAFM
mode, and the knob at the rear of the head adjusts AFM/LFM mode.
The range is -9.9 V to 9.9 V in both modes. The number will change
slowly at the extremes of the range and quickly around 0 V. Ideally,
the zeroed signal should be between ±0.1 V. Do this first in TMAFM
mode, then switch to AFM/LFM mode and try to zero the detector.
Flip back and forth between the two modes a few times (adjusting
each time) until the value in both modes is as close to 0 V as possible.
It will fluctuate during the experiment. If there is steady drift, you can
adjust it during the experiment. If the number won't settle down, the
laser could be at a bad position on the cantilever. Move the laser spot
and repeat (Figure 7.17). Always end this step in TMAFM mode.

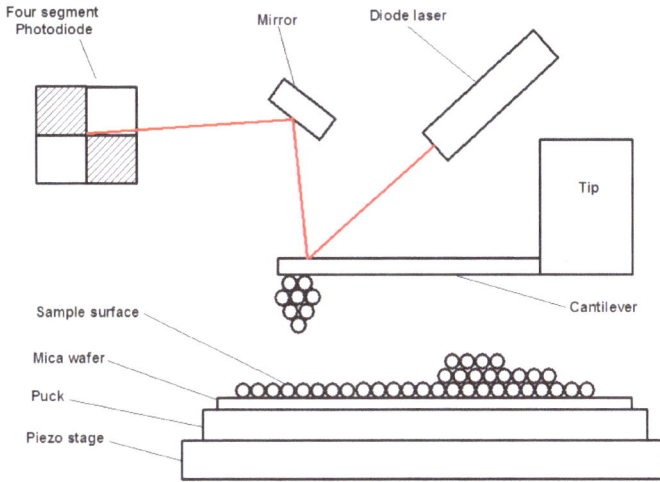

Figure 7.17: Schematic of the laser set-up.

12. Focus again on the sample surface.
13. The sample surface can still be moved with respect to the camera via the sample stage. In choosing a place to image nanoparticles, avoid anything that you can see on the sample surface. The scale on the screen is 18 μm per cm.

Tip tuning

1. Log onto computer.
2. The software is called Nanoscope. Close the version dialog box. Typically, the screen on the left will allow adjustment of software parameters, and the screen on the right will show the data.
3. On the adjustment screen, the two icons are to adjust the microscope (a picture of a microscope) and to perform data analysis (a picture of a rainbow). Click the microscope icon.
4. Under the microscope pull down menu, choose profile and select tapping AFM. Don't use another users profile. Use the "tapping" AFM.
5. Before beginning tapping mode, the cantilever must be tuned to ensure correct operation. Each tip has its own resonance frequency. The cantilever can be blindly auto tuned or manually tuned. However, the auto-tuning scheme can drive the amplitude so high as to damage the tip.

Auto-tuning

1. Click on the cantilever tune icon.
2. Click the auto-tune button. The computer will enter the tuning procedure, automatically entering such parameters as set point and drive amplitude. If tuned correctly, the drive frequency will be approximately 300 Hz.

Manually tuning

1. Click on the cantilever tune icon.
2. Select manual tuning under the sweep controls menu.
3. The plot is of amplitude (white) and phase (yellow) versus frequency. The sweep width is the x-range. The central frequency is the driving frequency which should be between 270-310 Hz. Typically, the initial plot will not show any peaks, and the x and y settings will need to be adjusted in order to see the resonance plot.
4. Widen the spectral window to about 100 Hz. The $270 - 310$ Hz window where the driving frequency will be set needs to be visible.
5. To zoom in use the green line (this software is not click and drag!):
   i. Left click separation.
   ii. Left click position.
   iii. Right click to do something.
   iv. Right click to clear lines.
6. If a peak is clipped, change the drive amplitude. Ideally this will be between 15 and 20 mV and should be below 500 mV. If a white line is not visible (there should be a white line along the bottom of the graph), the drive amplitude must be increased.
7. Ideally the peak will have a regular shape and only small shoulders. If there is a lot of noise, re-install the tip and things could improve. (Be careful as the auto-tuning scheme can drive the amplitude so high as to damage the tip.)
8. At this point, auto-tuning is okay. We can see that the parameters are reasonable. To continue the manual process, continue following these steps.
9. Adjust the drive amplitude so that the peak is at 2.0 V.
10. Amplitude set point while tuning corresponds to the vertical off set. If it is set to 0, the green line is 0.
11. Position the drive frequency not at the center of the peak, but instead at 5% toward the low energy (left) of the peak value. This offset is about $^4/_{10}{}^{th}$ of a division. Right click three times to execute this

change. This accounts for the damping that occurs when the tip approaches the sample surface.

12. Left monitor - channel 2 dialogue box - click zero phase.

Image acquisition

1. Click the eyeball icon for image mode.
2. Parameter adjustments
   i. Other controls.
   ii. Microscope mode: tapping.
   iii. Z-limit max height: 5.064 μm. This can be reduced if limited in Z-resolution.
   iv. Color table: 2.
   v. Engage set point: 1.00.
   vi. Serial number of this scanner (double check since this has the factory parameter and is different from the other AFM).
   vii. Parameter update retract; disabled.
3. Scan controls
   i. Scan size: 2 μm. Be careful when changing this value – it will automatically go between μm and nm (reasonable values are from 200 nm to 100 μm).
   ii. Aspect ratio: 1 to 1.
   iii. X and Y offset: 0.
   iv. Scan angle (like scan rotation): raster on the diagonal.
   v. Scan rate: 1.97 Hz is fast, and 100 Hz is slow.
4. Feedback control
   i. SPM: amplitude.
   ii. Integral gain: 0.5 (this parameter and the next parameter may be changed to improve image).
   iii. Proportional gain: 0.7.
   iv. Amplitude set point: 1 V.
   v. Drive frequency: from tuning.
   vi. Drive amplitude: from tuning.
5. Once all parameters are set, click engage (icon with green arrow down) to start engaging cantilever to sample surface and to begin image acquisition. The bottom of the screen should be "tip secured". When the tip reaches the surface it automatically begins imaging.
6. If the amplitude set point is high, the cantilever moves far away from the surface, since the oscillation is damped as it approaches. While in free oscillation (set amplitude set point to 3), adjust drive amplitude

so that the output voltage (seen on the scope) is 2 V. Big changes in this value while an experiment is running indicate that something is on the tip. Once the output voltage is at 2 V, bring the amplitude set point back down to a value that puts the z outer position line white and in the center of the bar on the software (1 V is very close).

7. Select channel 1 data type – height. Select channel 2 data type - amplitude. Amplitude looks like a 3D image and is an excellent visualization tool or for a presentation. However, the real data is the height data.

8. Bring the tip down (begin with amplitude set point to 2). The goal is to tap hard enough to get a good image, but not so hard as to damage the surface of the tip. Set to 3 clicks bellow jus touching by further lowering amplitude set point with 3 left arrow clicks on the keyboard. The tip z-center position scale on the right-hand screen shows the extension on the piezo scanner. When the tip is properly adjusted, expect this value to be near the center.

9. Select view/scope mode (the scope icon). Check to see if trace and retrace are tracking each other. If so, the lines should look the same, but they probably will not overlap each other vertically or horizontally. If they are tracking well, then your tip is scanning the sample surface and you may return to view/image mode (the image icon). If they are not tracking well, adjust the scan rate, gains, and/or set point to improve the tracking. If tracing and retrace look completely different, you may need to decrease the set point to improve the tracking. If trace and retrace look completely different, you may need to decrease the set point one or two clicks with the left arrow key until they start having common features in both directions. Then reduce the scan rate: a reasonable value for scan sizes of 1-3 μm would be 2 Hz. Next try increasing the integral gain. As you increase the integral gain, the tracking should improve, although you will reach a value beyond which the noise will increase as the feedback loop starts to oscillate. If this happens, reduce gains, if trace and retrace still do not track satisfactorily, reduce the set point again. Once the tip is tracking the surface, choose view/image mode.

10. Integral gain controls the amount of integrated error signal used in the feedback calculation. The higher this parameter is set, the better the tip will track the same topography. However, if it is set too high, noise due to feedback oscillation will be introduced into the scan.

11. Proportional gain controls the amount of proportional arrow signal used in the feedback calculation.

12. Once amplitude set point is adjusted with the phase data, change channel 2 to amplitude. The data scale can be changed (it is the same as for display as it does not affect the data). In the amplitude image, lowering the voltage increases the contrast.
13. Move small amounts on the image surface with x and y offset to avoid large, uninteresting objects. For example, setting the y offset to -2 will remove features at the bottom of the image, thus shifting the image up. Changing it to -3 will then move the image one more unit up. Make sure you are using μm and not nm if you expect to see a real change.
14. To move further, disengage the tip (click the red up arrow icon so that the tip moves up 25 μm and secures). Move upper translational stage to keep the tip in view in the light camera. Re-engage the tip.
15. If the shadow in the image is drawn out, the amplitude set point should be lowered even further. The area on the image that is being drawn is controlled by the frame pull-down menu (and the up and down arrows). Lower the set point and redraw the same neighborhood to see if there is improvement. The proportional and integral gain can also be adjusted.
16. The frame window allows you to restart from the top, bottom, or a particular line.
17. Another way to adjust the amplitude set point value is to click on signal scope to ensure trace and retrace overlap. To stop y rastering, slow scan axis.
18. To take a better image, increase the number of lines (512 is max), decrease the speed (1 Hz), and lower the amplitude set point. The resolution is about 10 nm in the x and y directions due to the size of the tip. The resolution in the z direction is less than 1 nm.
19. Changing the scan size allows us to zoom in on features. You can zoom in on a center point by using zoom in box (left clicking to toggle between box position and size), or you can manually enter a scan size on the left-hand screen.
20. Click on capture (the camera icon) to grab images. To speed things up, restart the scan at an edge to grab a new image after making any changes in the scan and feedback parameters. When parameters are changed, the capture option will toggle to "next". There is a forced capture option, which allows you to collect an image even if parameters have been switched during the capture. It is not completely reliable.

21. To change the file name, select capture filename under the capture menu. The file will be saved in the/directory which is *d:\capture*. To save the picture, under the utility pull-down menu select TIFF export. The zip drive is *G:*.

Image analysis

1. Analysis involves flattening the image and measuring various particle dimensions, click the spectrum button.
2. Select the height data (image pull-down menu, select left or right image). The new icons in the "analysis" menu are:
   - i.   Thumbnails.
   - ii.   Top view.
   - iii. Side view.
   - iv.   Section analysis.
   - v.   Roughness.
   - vi.   Rolling pin (flattening).
   - vii. Plane auto-fit.
3. To remove the bands (striping) in the image, select the rolling pin. The order of flattening is the order of the baseline correction. A raw offset is 0 order, a straight sloping line is order 1. Typically, a second order correction is chosen to remove "scanner bow" which are the dark troughs on the image plane.
4. To remove more shadows, draw exclusion boxes over large objects and then re-flatten. Be sure to save the file under a new name. The default is t overwrite it.
5. In section analysis, use the multiple cursor option to measure a particle in all dimensions. Select fixed cursor. You can save pictures of this information, but things must be written down! There is also a particle analysis menu.
6. Disengage the cantilever and make sure that the cantilever is in secure mode before you move the cantilever to the other spots or change to another sample.
7. Loosen the clamp to remove the tip and holder.
8. Remove the tip and replace it onto the gel sticky tape using the fine tweezers.
9. Recover the sample with tweezers.
10. Close the program.
11. Log out of the instrument.

12. After the experiment, turn off the monitor and the power of the light source. Leave the controller on.

Bibliography

R. Anderson and A. R. Barron, Reaction of hydroxyfullerene with metal salts: a route to remediation and immobilization. *J. Am. Chem. Soc.*, 2005, **127**, 10458.

M. Bellucci, G. Gaggiotti, M. Marchetti, F. Micciulla, R. Mucciato, and M. Regi, Atomic force microscopy characterization of carbon nanotubes. *J. Physics: Conference Series*, 2007, **61**, 99.

I. I. Bobrinetskii, V. N. Kukin, V. K. Nevolin, and M. M. Simunin, Carbon nanomaterial studied by atomic-force and electron microscopies. *Semiconductor*, 2008, **42**, 1496.

S. H. Cohen and M. L. Lightbody. *Atomic Force Microscopy/Scanning Tunneling Microscopy 2*. Plenum, New York (1997).

E. P. Dillon, C. A. Crouse, and A. R. Barron, Synthesis, characterization, and carbon dioxide adsorption of covalently attached polyethyleneimine-functionalized single-wall carbon nanotubes. *ACS Nano*, 2008, **2**, 156.

P. Egberts, G. H. Han, X. Z. Liu, A. T. C. Johnson, and R. W. Carpick, Frictional behavior of atomically thin sheets: hexagonal-shaped graphene islands grown on copper by chemical vapor deposition. *ACS Nano*, 2014, **8**, 5010.

P. Gong, Z. Ye, L. Yuan, and P. Egberts, Evaluation of wetting transparency and surface energy of pristine and aged graphene through nanoscale friction. *Carbon*, 2018, **132**, 749.

C. Gu, C. Ray, S. Guo, and B. B. Akhremitchev, Single-molecule force spectroscopy measurements of interactions between C_{60} fullerene molecules. *J. Phys. Chem.*, 2007, **111**, 35.

G. Kaupp, *Atomic Force Microscopy, Scanning Nearfield Optical Microscopy and Nanoscratching: Application to Rough and Natural Surfaces*. Springer-Verlag, Berlin (2006).

H. Lee, J. H. Ko, J. S. Choi, J. H. Hwang, Y. H. Kim, M. Salmeron, and J. Y. Park, Enhancement of friction by water intercalated between graphene and mica. *J. Phys. Chem. Lett.*, 2017, **8**, 3482.

C. Lee, Q. Li, W. Kalb, X. Z. Liu, H. Berger, R. W. Carpick, and J. Hone. Frictional characteristics of atomically thin sheets. *Science*, 2010, **328**, 76.

Q. Li, K. S. Kim, and A. Rydberg, Lateral force calibration of an atomic force microscope with a diamagnetic levitation spring system. *Rev. Sci. Instrum.*, 2006, **77**, 065105.

Q. Li, C. Lee, R. W. Carpick, and J. Hone, Substrate effect on thickness-dependent friction on graphene. *Phys. Status Solidi B*, 2010, **247**, 2909.

S. Morita, R. Wiesendanger, E. Meyer, and F. J. Giessibl. *Noncontact Atomic Force Microscopy*. Springer, Berlin (2002).

Z. Ye, P. Egberts, G. H. Han, A. T. C. Johnson, R. W. Carpick, and A. Martini, Load-dependent friction hysteresis on graphene. *ACS Nano*, 2016, **10**, 5161.

Chapter 8: Magnetic Force Microscopy

Samantha L. Donaldson, Pavan M. V. Raja and Andrew R. Barron

Introduction

Magnetic force microscopy (MFM) is a natural extension of scanning tunneling microscopy (STM), whereby both the physical topology of a sample surface and the magnetic topology may be seen. Magnetic force microscopy was not far behind, with the first report of its use in 1987 by Yves Martin and H. Kumar Wickramasinghe. An AFM with a magnetic tip was used to perform these early experiments, which proved to be useful in imaging both static and dynamic magnetic fields.

MFM, AFM, and STM all have similar instrumental setups, all of which are based on the early scanning tunneling microscopes. In essence, STM uses a very small conductive tip attached to a piezoelectric cylinder to carefully scan across a small sample space. The electrostatic forces between the conducting sample and tip are measured, and the output is a picture that shows the surface of the sample. AFM and MFM are essentially derivative types of STM, which explains why a typical MFM device is very similar to an STM, with a piezoelectric driver and magnetized tip as seen in Figure 8.1.

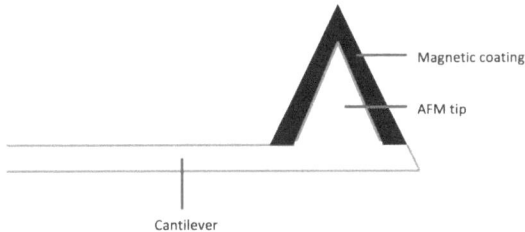

Figure 8.1: Illustration of an MFM tip on the instrument cantilever.

The MFM instrument very closely resembles an atomic force microscope, and this is for good reason. The simplest MFM instruments are no more than AFM instruments with a magnetic tip. The differences between AFM and MFM lie in the data collected and its processing. Where AFM gives topological data through tapping, noncontact, or contact mode, MFM gives both topological (tapping) and magnetic topological (non-contact) data through a two-scan process known as interleave scanning. The relationships between basic STM, AFM, and MFM are summarized in Table 8.1.

Technique	Samples	Qualities observed	Modes	Benefits	Limitations
MFM	Any film or powder sur- face; magnetic	Electrostatic interactions; magnetic forces/domains; van der Waals' interactions; topology; morphology	Tapping; non- contact	Magnetic and physical properties; high resolution	Resolution depends on tip size; different tips for various applications; complicated data processing and analysis
STM	Only conductive surfaces	Topology; morphology	Constant height; constant current	Simplest instrumental setup; many variations	Resolution depends on tip size; tips wear out easily; rare technique
AFM	Any film or powder surface	Particle size; topology; morphology	Tapping; contact; non-contact	Common, standardized; often do not need special tip; ease of data analysis	Resolution depends on tip size; easy to break tips; slow process

Table 8.1: A summary of the capabilities of MFM, SPM, and AFM instrumentation.

Data collection

Interleave scanning, also known as two-pass scanning, is a process typically used in an MFM experiment. The magnetized tip is first passed across the sample in tapping mode, similar to an AFM experiment, and this gives the surface topology of the sample. Then, a second scan is taken in non-contact mode, where the magnetic force exerted on the tip by the sample is measured. These two types of scans are shown in Figure 8.2.

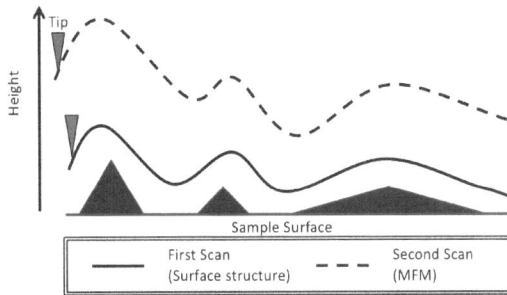

Figure 8.2: Interleave (two-pass) scanning across a sample surface.

In non-contact mode (also called dynamic or AC mode), the magnetic force gradient from the sample affects the resonance frequency of the MFM cantilever and can be measured in three different ways.

- Phase detection: the phase difference between the oscillation of the cantilever and piezoelectric source is measured
- Amplitude detection: the changes in the cantilever's oscillations are measured
- Frequency modulation: the piezoelectric source's oscillation frequency is changed to maintain a 90° phase lag between the cantilever and the piezoelectric actuator. The frequency change needed for the lag is measured.

Regardless of the method used in determining the magnetic force gradient from the sample, a MFM interleave scan will always give the user information about both the surface and magnetic topology of the sample. A typical sample size is 100×100 μm, and the entire sample is scanned by rastering from one line to another. In this way, the MFM data processor can compose an image of the surface by combining lines of data from either the surface or magnetic scan. The output of an MFM scan is two images, one showing the surface and the other showing magnetic qualities of the sample. An idealized example is shown in Figure 8.3.

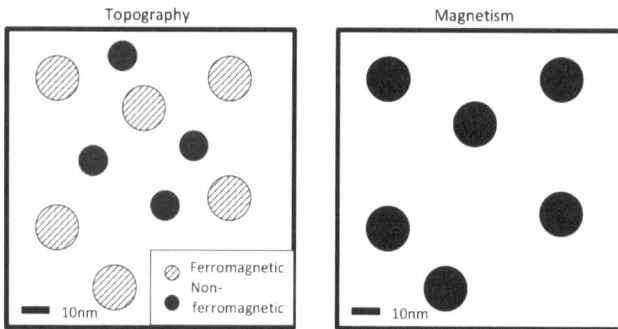

Figure 8.3: Idealized images of a mixture of ferromagnetic and non-ferromagnetic nanoparticles from MFM.

Types of MFM tips

Any suitable magnetic material or coating can be used to make an MFM tip. Some of the most commonly used standard tips are coated with FeNi, CoCr, and NiP, while many research applications call for individualized tips such as carbon nanotubes. The resolution of the end image in MFM is dependent directly on the size of the tip, therefore MFM tips must come to a sharp point on the angstrom scale in order to function at high resolution. This leads to tips being costly, an issue exacerbated by the fact that coatings are often soft or brittle, leading to wear and tear. The best materials for MFM tips, therefore, depend on the desired resolution and application. For example, a high coercivity coating such as CoCr may be favored for analyzing bulk or strongly magnetic samples, whereas a low coercivity material such as FeNi might be preferred for more fine and sensitive applications.

Data output and applications

From an MFM scan, the product is a 2D scan of the sample surface, whether this be the physical or magnetic topographical image. Importantly, the resolution depends on the size of the tip of the probe; the smaller the probe, the higher the number of data points per square micrometer and therefore the resolution of the resulting image. MFM can be extremely useful in determining the properties of new materials, as in Figure 8.4, or in analyzing already known materials' magnetic landscapes. This makes MFM particularly useful for the analysis of hard drives. As people store more and more information on magnetic storage devices, higher storage capacities need to be developed and emergency backup procedures for this data must be developed. MFM is an ideal procedure for characterizing the fine magnetic surfaces of hard drives for use in research and development, and also can show the magnetic surfaces of already-used hard drives for data recovery in the event of a hard drive malfunction. This is useful both in forensics and in researching new magnetic storage materials.

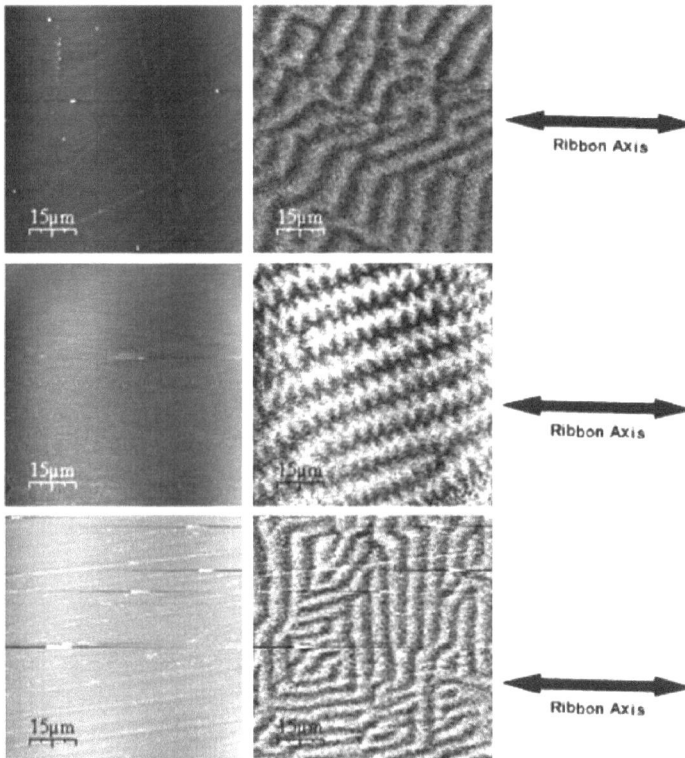

Figure 8.4: Images of $Fe_{40}Ni_{38}Mo_4B_{18}$ ribbons from MFM. Left images: surface topography. Right images: magnetic topography. Reproduced with permission from I. García, N. Iturriza, J. José del Val, H. Grande, J. A. Pomposo, and J. González, Magnetic force microscopy characterization of heat and current treated $Fe_{40}Ni_{38}Mo_4B_{18}$ amorphous ribbons. *J. Magn. Magn. Mater.*, 2010, 13, 1822. Copyright: Elsevier (2010).

MFM has also found applications on the frontiers of research, most notably in the field of Spintronics. In general, Spintronics is the study of the spin and magnetic moment of solid-state materials, and the manipulation of these properties to create novel electronic devices. One example of this is quantum computing, which is promising as a fast and efficient alternative to traditional transistor-based computing. With regards to Spintronics, MFM can be used to characterize non-homogenous magnetic materials and unique samples such as dilute magnetic semiconductors (DMS). This is useful for research in magnetic storage such as MRAM, semiconductors, and magnetoresistive materials.

MFM for characterization of magnetic storage devices

In device manufacturing, the smoothness and/or roughness of the magnetic coatings of hard drive disks are significant in their ability to operate. Smoother coatings provide a low magnetic noise level, but stick to read/write heads, whereas rough surfaces have the opposite qualities. Therefore, fine tuning not only of the magnetic properties but the surface qualities of a given magnetic film is extremely important in the development of new hard drive technology. Magnetic force microscopy allows the manufacturers of hard drives to analyze disks for magnetic and surface topology, making it easier to control the quality of drives and determine which materials are suitable for further research. Industrial competition for higher bit density (bits per square millimeter), which means faster processing and increased storage capability, means that MFM is very important for characterizing films to very high resolution.

Conclusion

Magnetic force microscopy is a powerful surface technique used to deduce both the magnetic and surface topology of a given sample. In general, MFM offers high resolution, which depends on the size of the tip, and straightforward data once processed. The images outputted by the MFM raster scan are clear and show structural and magnetic features of a 100×100 μm square of the given sample. This information can be used not only to examine surface properties, morphology, and particle size, but also to determine the bit density of hard drives, features of magnetic computing materials, and identify exotic magnetic phenomena at the atomic level. As MFM evolves, thinner and thinner magnetic tips are being fabricated to finer applications, such as in the use of carbon nanotubes as tips to give high atomic resolution in MFM images. The customizability of magnetic coatings and tips, as well as the use of AFM equipment for MFM, make MFM an important technique in the electronics industry, making it possible to see magnetic domains and structures that otherwise would remain hidden.

Bibliography

I. García, N. Iturriza, J. José del Val, H. Grande, J. A. Pomposo, and J. González, Magnetic force microscopy characterization of heat and current treated $Fe_{40}Ni_{38}Mo_4B_{18}$ amorphous ribbons. *J. Magn. Magn. Mater.,* 2010, **13**, 1822.

Y. Martin and H. K. Wickramasinghe, Magnetic imaging by ''force microscopy'' with 1000 Å resolution. *Appl. Phys. Lett.* 1987, **50**, 1455.

G. Persch and H. Strecker, Applications of magnetic force microscopy in magnetic storage device manufacturing. *Ultramicroscopy,* 1992, **42-44,** 1269.

S. Salapaka and M. Salapaka, Scanning probe microscopy. *IEEE Control Syst. Mag.,* 2008, **2**, 65.

K. Tanaka, M. Yoshimura, and K. Ueda, High-resolution magnetic force microscopy using carbon nanotube probes fabricated directly by microwave plasma-enhanced chemical vapor deposition. *J. Nanomater.,* 2009, 2009, 147204.

I. Žutić, J. Fabian, and S. Das Sarma, Spintronics: fundamentals and applications. *Rev. Mod. Phys.,* 2004, **2**, 323.

Chapter 9: UV-Visible Spectroscopy

Yongji Gong, Sravani Gullapalli,
Nicole Moody, Brittany L. Oliva-Chatelain, Eric Wagner,
Pavan M. V. Raja and Andrew R. Barron

Introduction

Absorption spectroscopy, in general, refers to characterization techniques that measure the absorption of radiation by a material, as a function of the wavelength. Depending on the source of light used, absorption spectroscopy can be broadly divided into infrared and UV-visible spectroscopy. The band gap of Group 12-16 semiconductors is in the UV-visible region. This means the minimum energy required to excite an electron from the valence states of the Group 12-16 semiconductor QDs to its conduction states, lies in the UV-visible region. This is also a reason why most of the Group 12-16 semiconductor quantum dot solutions are colored.

This technique is complementary to fluorescence spectroscopy, in that UV-visible spectroscopy measures electronic transitions from the ground state to the excited state, whereas fluorescence deals with the transitions from the excited state to the ground state. In order to characterize the optical properties of a quantum dot, it is important to characterize the sample with both these techniques

In quantum dots, due to the very small number of atoms, the addition or removal of one atom to the molecule changes the electronic structure of the quantum dot dramatically. Taking advantage of this property in Group 12-16 semiconductor quantum dots, it is possible to change the band gap of the material by just changing the size of the quantum dot. A quantum dot can absorb energy in the form of light over a range of wavelengths, to excite an electron from the ground state to its excited state. The minimum energy that is required to excite an electron, is dependent on the band gap of the quantum dot. Thus, by making accurate measurements of light absorption at different wavelengths in the ultraviolet and visible spectrum, a correlation can be made between the band gap and size of the quantum dot. Group 12-16 semiconductor quantum dots are of particular interest, since their band gap lies in the visible region of the solar spectrum.

The UV-visible absorbance spectroscopy is a characterization technique in which the absorbance of the material is studied as a function of wavelength. The visible region of the spectrum is in the wavelength range of 380 nm (violet) to 740 nm (red) and the near ultraviolet region extends to wavelengths of about 200 nm. The UV-visible spectrophotometer analyzes over the wavelength range 200 – 900 nm.

When the Group 12-16 semiconductor nanocrystals are exposed to light having an energy that matches a possible electronic transition as dictated by laws of quantum physics, the light is absorbed, and an exciton pair is formed. The UV-visible spectrophotometer records the wavelength at which the absorption occurs along with the intensity of the absorption at each wavelength. This is recorded in a graph of absorbance of the nanocrystal versus wavelength.

Instrumentation

A working schematic of the UV-visible spectrophotometer is shown in Figure 9.1.

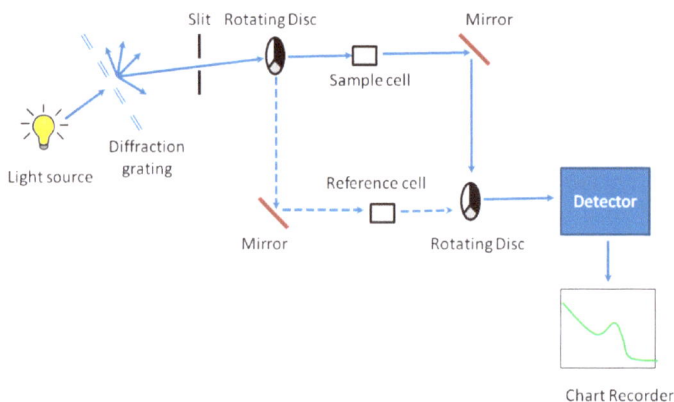

Figure 9.1: Schematic of UV-visible spectrophotometer.

The light source

Since it is a UV-vis spectrophotometer, the light source (Figure 9.1) needs to cover the entire visible and the near ultra-violet region (200 - 900 nm). Since it is not possible to get this range of wavelengths from a single lamp, a combination of a deuterium lamp for the UV region of the spectrum and tungsten or halogen lamp for the visible region is used. This output is then sent through a diffraction grating as shown in the schematic.

The diffraction grating and the slit

The beam of light from the visible and/or UV light source is then separated into its component wavelengths (like a very efficient prism) by a diffraction grating (Figure 9.1). Following the slit is a slit that sends a monochromatic beam into the next section of the spectrophotometer.

Rotating discs

Light from the slit then falls onto a rotating disc (Figure 9.1). Each disc consists of different segments – an opaque black section, a transparent section and a mirrored section. If the light hits the transparent section, it will go straight through the sample cell, get reflected by a mirror, hits the mirrored section of a second rotating disc, and then collected by the detector. Else if the light hits the mirrored section, gets reflected by a mirror, passes through the reference cell, hits the transparent section of a second rotating disc and then collected by the detector. Finally, if the light hits the black opaque section, it is blocked and no light passes through the instrument, thus enabling the system to make corrections for any current generated by the detector in the absence of light.

Sample cell, reference cell and sample preparation

For liquid samples, a square cross section tube sealed at one end is used. The choice of cuvette depends on the following factors:

- **Type of solvent:** For aqueous samples, specially designed rectangular quartz, glass or plastic cuvettes are used. For organic samples glass and quartz cuvettes are used.
- **Excitation wavelength:** Depending on the size and thus, bandgap of the 12-16 semiconductor nanoparticles, different excitation wavelengths of light are used. Depending on the excitation wavelength, different materials are used (Table 9.1).
- **Cost:** Plastic cuvettes are the least expensive and can be discarded after use. Though quartz cuvettes have the maximum utility, they are the most expensive, and need to be reused. Generally, disposable plastic cuvettes are used when speed is more important than high accuracy.

Cuvette	Wavelength (nm)
Visible only glass	380 - 780
Visible only plastic	380 - 780
UV plastic	220 - 780
Quartz	200 - 900

Table 9.1: Cuvette materials and their wavelengths.

The best cuvettes need to be very clear and have no impurities that might affect the spectroscopic reading. Defects on the cuvette such as scratches, can scatter light and hence should be avoided. Some cuvettes are clear only on two sides and can be used in the UV-visible spectrophotometer but cannot be used for fluorescence spectroscopy measurements. For Group 12-16 semiconductor nanoparticles prepared in organic solvents, the quartz cuvette is chosen.

In the sample cell the quantum dots are dispersed in a solvent, whereas in the reference cell the pure solvent is taken. It is important that the sample be very dilute (maximum first exciton absorbance should not exceed 1 au) and the solvent is not UV-visible active. For these measurements, it is required that the solvent does not have characteristic absorption or emission in the region of interest. Solution phase experiments are preferred, though it is possible to measure the spectra in the solid state also using thin films, powders, etc. The instrumentation for solid state UV-visible absorption spectroscopy is slightly different from the solution phase experiments and is beyond the scope of discussion.

Detector

Detector converts the light into a current signal that is read by a computer. Higher the current signal, greater is the intensity of the light. The computer then calculates the absorbance using the in,

$$A = \log_{10}(I_0/I)$$

where A denotes absorbance, I is sample cell intensity, and I_o is the reference cell intensity.

The following cases are possible:

- Where $I < I_0$ and $A < 0$. This usually occurs when the solvent absorbs in the wavelength range. Preferably the solvent should be changed, to get an accurate reading for actual reference cell intensity.
- Where $I = I_0$ and $A = 0$. This occurs when pure solvent is put in both reference and sample cells. This test should always be done before testing the sample, to check for the cleanliness of the cuvettes.
- When $A = 1$. This occurs when 90% or the light at a particular wavelength has been absorbed, which means that only 10% is seen at the detector. So I_0/I becomes $100/10 = 10$. Log_{10} of 10 is 1.
- When $A > 1$. This occurs in extreme case where more than 90% of the light is absorbed.

Output

The output is the form of a plot of absorbance against wavelength, e.g., Figure 9.2.

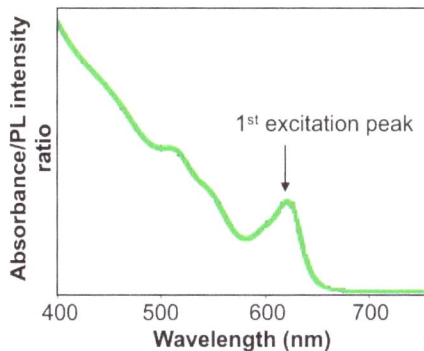

Figure 9.2: Representative UV-visible absorption spectrum for CdSe tetrapods.

Beer-Lambert law

In order to make comparisons between different samples, it is important that all the factors affecting absorbance should be constant except the sample itself.

Effect of concentration on absorbance

The extent of absorption depends on the number of absorbing nanoparticles or in other words the concentration of the sample. If it is a reasonably concentrated solution, it will have a high absorbance since there are lots of nanoparticles to interact with the light. Similarly, in an extremely dilute solution, the absorbance is very low. In order to compare two solutions, it is important that we should make some allowance for the concentration.

Effect of container shape

Even if we had the same concentration of solutions, if we compare two solutions, one in a rectangular shaped container (e.g., Figure 9.3) so that light travelled 1 cm through it and the other in which the light travelled 100 cm through it, the absorbance would be different. This is because if the length the light travelled is greater, it means that the light interacted with a greater number of nanocrystals, and thus has a higher absorbance. Again, in order to compare two solutions, it is important that we should make some allowance for the concentration.

Figure 9.3: A typical rectangular cuvette for UV-visible spectroscopy.

The law

The Beer-Lambert law addresses the effect of concentration and container shape as shown in,

$$\log_{10}(I_0/I) = \varepsilon l c$$

and

$$A = \varepsilon l c$$

where A denotes absorbance; ε is the molar absorptivity or molar absorption coefficient; l is the path length of light (in cm); and c is the concentration of the solution (mol/dm^3).

Molar absorptivity

From the Beer-Lambert law, the molar absorptivity 'ε' can be expressed as shown in,

$$c = A/l\varepsilon$$

Molar absorptivity corrects for the variation in concentration and length of the solution that the light passes through. It is the value of absorbance when light passes through 1 cm of a 1 mol/dm^3 solution.

Limitations of Beer-Lambert law

The linearity of the Beer-Lambert law is limited by chemical and instrumental factors.

- At high concentrations (> 0.01 M), the relation between absorptivity coefficient and absorbance is no longer linear. This is due to the electrostatic interactions between the quantum dots in close proximity.
- If the concentration of the solution is high, another effect that is seen is the scattering of light from the large number of quantum dots.
- The spectrophotometer performs calculations assuming that the refractive index of the solvent does not change significantly with the presence of the quantum dots. This assumption only works at low concentrations of the analyte (quantum dots).
- Presence of stray light.

Semiconductor quantum nanoparticles

In the case of semiconductor nanocrystals, the effect of the size on the optical properties of the particles is very interesting. Consider a Group 12-16 semiconductor, cadmium selenide (CdSe). A 2 nm sized CdSe crystal has a blue color fluorescence whereas a larger nanocrystal of CdSe of about 6 nm has a dark red fluorescence (Figure 9.4). In order to understand the size dependent optical properties of semiconductor nanoparticles, it is important to know the physics behind what is happening at the nano level.

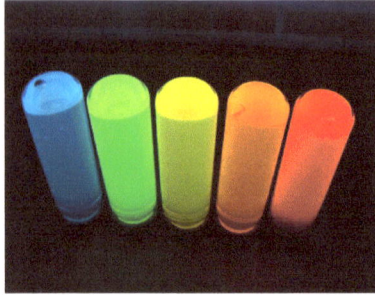

Figure 9.4: Fluorescing CdSe quantum dots synthesized in a heat transfer liquid of different sizes (M. S. Wong, Rice University).

Energy levels in a semiconductor

The electronic structure of any material is given by a solution of Schrödinger equations with boundary conditions, depending on the physical situation. The electronic structure of a semiconductor (Figure 9.5) can be described by the following terms:

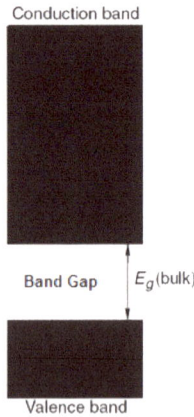

Figure 9.5: Simplified representation of the energy levels in a bulk semiconductor.

Energy level

By the solution of Schrödinger's equations, the electrons in a semiconductor can have only certain allowable energies, which are associated with energy levels. No electrons can exist in between these levels, or in other words can have energies in between the allowed energies. In addition, from Pauli's Exclusion Principle, only 2 electrons with opposite spin can exist at any one energy level. Thus, the electrons start filling from the lowest energy levels. Greater the number of atoms in a crystal, the difference in allowable energies become very small, thus the distance between energy levels decreases. However, this distance can never be zero. For a bulk semiconductor, due to the large number of atoms, the distance between energy levels is very small and for all practical purpose the energy levels can be described as continuous (Figure 9.5).

Band gap

From the solution of Schrödinger's equations, there are a set of energies which is not allowable, and thus no energy levels can exist in this region. This region is called the band gap and is a quantum mechanical phenomenon (Figure 9.5). In a bulk semiconductor the bandgap is fixed; whereas in a quantum dot nanoparticle the bandgap varies with the size of the nanoparticle.

Valence band

In bulk semiconductors, since the energy levels can be considered as continuous, they are also termed as energy bands. Valence band contains electrons from the lowest energy level to the energy level at the lower edge of the bandgap (Figure 9.5). Since filling of energy is from the lowest energy level, this band is usually almost full.

Conduction band

The conduction band consists of energy levels from the upper edge of the bandgap and higher (Figure 9.5). To reach the conduction band, the electrons in the valence band should have enough energy to cross the band gap. Once the electrons are excited, they subsequently relax back to the valence band (either radiatively or non-radiatively) followed by a subsequent emission of radiation. This property is responsible for most of the applications of quantum dots.

Exciton and exciton Bohr radius

When an electron is excited from the valence band to the conduction band, corresponding to the electron in the conduction band a hole (absence of electron) is formed in the valence band. This electron pair is called an exciton. Excitons have a natural separation distance between the electron and hole, which is characteristic of the material. This average distance is called exciton

Bohr radius. In a bulk semiconductor, the size of the crystal is much larger than the exciton Bohr radius and hence the exciton is free to move throughout the crystal.

Energy levels in a quantum dot semiconductor

Before understanding the electronic structure of a quantum dot semiconductor, it is important to understand what a quantum dot nanoparticle is. We earlier studied that a nanoparticle is any particle with one of its dimensions in the 1 - 100 nm. A quantum dot is a nanoparticle with its diameter on the order of the materials exciton Bohr radius. Quantum dots are typically 2 - 10 nm wide and approximately consist of 10 to 50 atoms. With this understanding of a quantum dot semiconductor, the electronic structure of a quantum dot semiconductor can be described by the following terms (Figure 9.6).

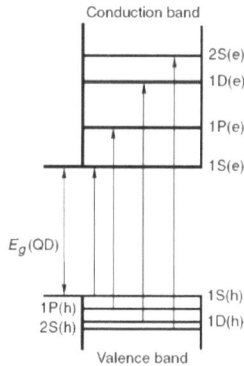

Figure 9.6: Energy levels in quantum dot. Allowed optical transitions are shown. Adapted from T. Pradeep, *Nano: The Essentials. Understanding Nanoscience and Nanotechnology*, Tata McGraw-Hill, New Delhi (2007).

Quantum confinement

When the size of the semiconductor crystal becomes comparable or smaller than the exciton Bohr radius, the quantum dots are in a state of quantum confinement. As a result of quantum confinement, the energy levels in a quantum dot are discrete (Figure 9.6) as opposed to being continuous in a bulk crystal (Figure 9.5).

Discrete energy levels

In materials that have small number of atoms and are considered as quantum confined, the energy levels are separated by an appreciable amount of energy such that they are not continuous but are discrete (see Figure 9.6). The energy associated with an electron (equivalent to conduction band energy level) is given by is given by,

$$E^e = \frac{h^2 n^2}{8\pi^2 m_e d^2}$$

where h is the Planck's constant, m_e is the effective mass of electron and n is the quantum number for the conduction band states, and n can take the values 1, 2, 3 and so on. Similarly, the energy associated with the hole (equivalent to valence band energy level) is given by,

$$E^h = \frac{h^2 n'^2}{8\pi^2 m_h d^2}$$

where n' is the quantum number for the valence states, and n' can take the values 1, 2, 3, and so on. The energy increases as one goes higher in the quantum number. Since the electron mass is much smaller than that of the hole, the electron levels are separated more widely than the hole levels.

Tunable band gap

The energy levels are affected by the diameter of the semiconductor particles. If the diameter is very small, since the energy is dependent on inverse of diameter squared, the energy levels of the upper edge of the band gap (lowest conduction band level) and lower edge of the band gap (highest valence band level) change significantly with the diameter of the particle and the effective mass of the electron and the hole, resulting in a size dependent tunable band gap. This also results in the discretization of the energy levels.

Qualitatively, this can be understood in the following way. In a bulk semiconductor, the addition or removal of an atom is insignificant compared to the size of the bulk semiconductor, which consists of a large number of atoms. The large size of bulk semiconductors makes the changes in band gap so negligible on the addition of an atom, that it is considered as a fixed band gap. In a quantum dot, addition of an atom does make a difference, resulting in the tunability of band gap.

UV-visible absorbance

Due to the presence of discrete energy levels in a QD, there is a widening of the energy gap between the highest occupied electronic states and the lowest unoccupied states as compared to the bulk material. As a consequence, the optical properties of the semiconductor nanoparticles also become size dependent.

The minimum energy required to create an exciton is the defined by the band gap of the material, i.e., the energy required to excite an electron from the highest level of valence energy states to the lowest level of the conduction energy states. For a quantum dot, the bandgap varies with the size of the particle. It can be inferred that the band gap becomes higher as the particle becomes smaller. This means that for a smaller particle, the energy required for an electron to get excited is higher. The relation between energy and wavelength is given by,

$$E = \frac{hc}{\lambda}$$

where h is the Planck's constant, c is the speed of light, λ is the wavelength of light. Therefore, to cross a bandgap of greater energy, shorter wavelengths of light are absorbed, i.e., a blue shift is seen.

For Group 12-16 semiconductors, the bandgap energy falls in the UV-visible range. That is ultraviolet light or visible light can be used to excite an electron from the ground valence states to the excited conduction states. In a bulk semiconductor the band gap is fixed, and the energy states are continuous. This results in a rather uniform absorption spectrum (Figure 9.7a).

In the case of Group 12-16 quantum dots, since the bandgap can be changed with the size, these materials can absorb over a range of wavelengths. The peaks seen in the absorption spectrum (Figure 9.7b) correspond to the optical transitions between the electron and hole levels. The minimum energy and thus the maximum wavelength peak correspond to the first exciton peak or the energy for an electron to get excited from the highest valence state to the lowest conduction state. The quantum dot will not absorb wavelengths of energy longer than this wavelength. This is known as the absorption onset.

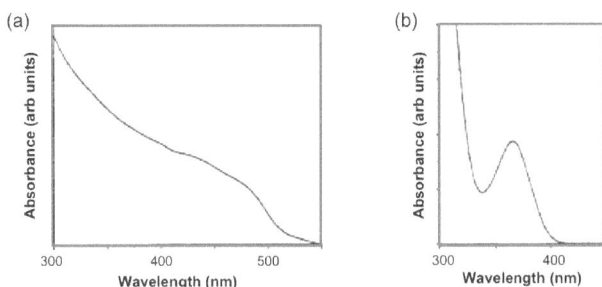

Figure 9.7: UV-vis spectra of (a) bulk CdS and (b) 4 nm CdS. Adapted from G. Kickelbick, *Hybrid Materials: Synthesis, Characterization and Applications*, Wiley-VCH, Weinheim (2007). Copyright: Wiley-VCH (2007).

Fluorescence

Fluorescence is the emission of electromagnetic radiation in the form of light by a material that has absorbed a photon. When a semiconductor quantum dot (QD) absorbs a photon/energy equal to or greater than its band gap, the electrons in the QD's get excited to the conduction state. This excited state is however not stable. The electron can relax back to its ground state by either emitting a photon or lose energy via heat losses. These processes can be divided into two categories – radiative decay and non-radiative decay. Radiative decay is the loss of energy through the emission of a photon or radiation. Non-radiative decay involves the loss of heat through lattice vibrations and this usually occurs when the energy difference between the levels is small. Non-radiative decay occurs much faster than radiative decay.

Usually the electron relaxes to the ground state through a combination of both radiative and non-radiative decays. The electron moves quickly through the conduction energy levels through small non-radiative decays and the final transition across the band gap is via a radiative decay. Large nonradiative decays don't occur across the band gap because the crystal structure can't withstand large vibrations without breaking the bonds of the crystal. Since some of the energy is lost through the non-radiative decay, the energy of the emitted photon, through the radiative decay, is much lesser than the absorbed energy. As a result, the wavelength of the emitted photon or fluorescence is longer than the wavelength of absorbed light. This energy difference is called the Stokes shift. Due this Stokes shift, the emission peak corresponding to the absorption band edge peak is shifted towards a higher wavelength (lower energy), i.e., Figure 9.8.

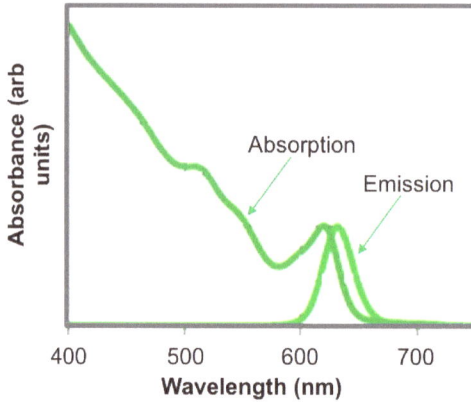

Figure 9.8: (a) Absorption spectra and (b) emission spectra of CdSe tetrapod.

Intensity of emission versus wavelength is a bell-shaped Gaussian curve. As long as the excitation wavelength is shorter than the absorption onset, the maximum emission wavelength is independent of the excitation wavelength. Figure 9.8 shows a combined absorption and emission spectrum for a typical CdSe tetrapod.

Analysis of data

The data obtained from the spectrophotometer is a plot of absorbance as a function of wavelength. Quantitative and qualitative data can be obtained by analyzing this information

Quantitative Information

The band gap of the semiconductor quantum dots can be tuned with the size of the particles. The minimum energy for an electron to get excited from the ground state is the energy to cross the band gap. In an absorption spectrum this is given by the first exciton peak at the maximum wavelength (λ_{max}).

Size of the quantum dots

The size of quantum dots can be approximated corresponding to the first exciton peak wavelength. Empirical relationships have been determined relating the diameter of the quantum dot to the wavelength of the first exciton peak. As an example, Group 12-16 semiconductor quantum dots cadmium selenide (CdSe), cadmium telluride (CdTe) and cadmium sulfide (CdS) were found to have the following empirical relationships as determined by fitting experimental data of absorbance versus wavelength of known sizes of particles,

$$D = (9.8127 \times 10^{-7})\lambda^3 - (1.7147 \times 10^{-3})\lambda^2 + (1.0064)\lambda - 194.84$$

$$D = (1.6122 \times 10^{-7})\lambda^3 - (2.6575 \times 10^{-6})\lambda^2 + (1.6242 \times 10^{-3})\lambda + 41.57$$

$$D = (-6.6521 \times 10^{-8})\lambda^3 + (1.9577 \times 10^{-4})\lambda^2 - (9.2352 \times 10^{-2})\lambda + 13.29$$

where D is the diameter and λ is the wavelength corresponding to the first exciton peak. For example, if the first exciton peak of a CdSe quantum dot is 500 nm, the corresponding diameter of the quantum dot is 2.345 nm and for a wavelength of 609 nm, the corresponding diameter is 5.008 nm.

Concentration of sample

Using the Beer-Lambert law, it is possible to calculate the concentration of the sample if the molar absorptivity for the sample is known. The molar absorptivity can be calculated by recording the absorbance of a standard solution of 1 mol/dm^3 concentration in a standard cuvette where the light travels a constant distance of 1 cm. Once the molar absorptivity and the absorbance of the sample are known, with the length the light travels being fixed, it is possible to determine the concentration of the sample solution.

Empirical equations can be determined by fitting experimental data of extinction coefficient per mole of Group 12-16 semiconductor quantum dots, at 250 °C, to the diameter of the quantum dot,

$$\varepsilon = 10043 \times D^{2.12}$$

$$\varepsilon = 5857 \times D^{2.65}$$

$$\varepsilon = 21536 \times D^{2.3}$$

The concentration of the quantum dots can then be then be determined by using the Beer Lambert law.

Qualitative Information

Apart from quantitative data such as the size of the quantum dots and concentration of the quantum dots, a lot of qualitative information can be derived from the absorption spectra.

Size distribution

If there is a very narrow size distribution, the first exciton peak will be very sharp (Figure 9.9). This is because due to the narrow size distribution, the differences in band gap between different sized particles will be very small and hence most of the electrons will get excited over a smaller range of wavelengths. In addition, if there is a narrow size distribution, the higher exciton peaks are also seen clearly.

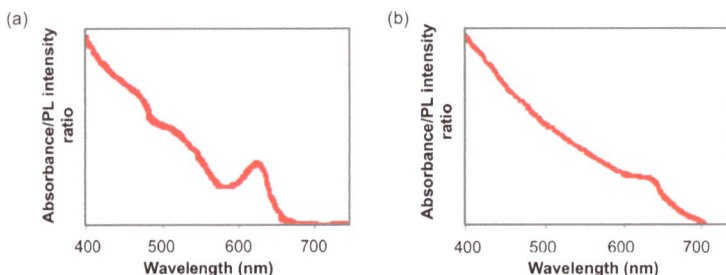

Figure 9.9: Examples of (a) narrow emission spectra and (b) broad emission spectra of CdSe QDs.

Shaped particles

In the case of a spherical quantum dot, in all dimensions, the particle is quantum confined (Figure 9.10). In the case of a nanorod, whose length is not in the quantum regime, the quantum effects are determined by the width of the nanorod. Similar is the case in tetrapods or four legged structures. The quantum effects are determined by the thickness of the arms. During the synthesis of the shaped particles, the thickness of the rod or the arm of the tetrapod does not vary among the different particles, as much as the length of the rods or arms changes. Since the thickness of the rod or tetrapod is responsible for the quantum effects, the absorption spectrum of rods and tetrapods have sharper features as compared to that of a quantum dot. Hence, qualitatively it is possible to differentiate between quantum dots and other shaped particles.

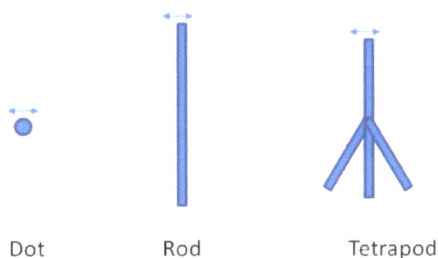

Dot   Rod   Tetrapod

Figure 9.10: Different shaped nanoparticles with the arrows indicating the dimension where quantum confinement effects are observed.

Crystal lattice information

In the case of CdSe semiconductor quantum dots it has been shown that it is possible to estimate the crystal lattice of the quantum dot from the adsorption spectrum (Figure 9.11), and hence determine if the structure is zinc blend or wurtzite.

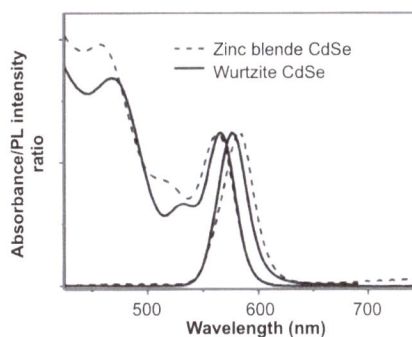

Figure 9.11: Zinc blende and wurtzite CdSe absorption spectra. Adapted from J. Jasieniak, C. Bullen, J. van Embden, and P. Mulvaney, Phosphine-free synthesis of CdSe nanocrystals. *J. Phys. Chem. B*, 2005, 109, 20665. Copyright: American Chemical Society (2005).

Factors affecting the optical properties of NPs

There are various factors that affect the absorption and emission spectra for semiconductor quantum crystals. Fluorescence is much more sensitive to the background, environment, presence of traps and the surface of the QDs than UV-visible absorption. Some of the major factors influencing the optical properties of quantum nanoparticles include:

Surface defects, imperfection of lattice, surface charges

The surface defects and imperfections in the lattice structure of semiconductor quantum dots occur in the form of unsatisfied valencies. Similar to surface charges, unsatisfied valencies provide a sink for the charge carriers, resulting in unwanted recombination.

Surface ligands

The presence of surface ligands is another factor that affects the optical properties. If the surface ligand coverage is a 100%, there is a smaller chance of surface recombination to occur.

Solvent polarity

The polarity of solvents is very important for the optical properties of the nanoparticles. If the quantum dots are prepared in organic solvent and have an organic surface ligand, the more non-polar the solvent, the particles are more dispersed. This reduces the loss of electrons through recombination again, since when particles come in close proximity to each other, increases the non-radiative decay events.

Characterization of silicon quantum dots

Silicon quantum dots are synthesized in inverse micelles. $SiCl_4$ is reduced using a two-fold excess of $LiAlH_4$ (Figure 9.12).

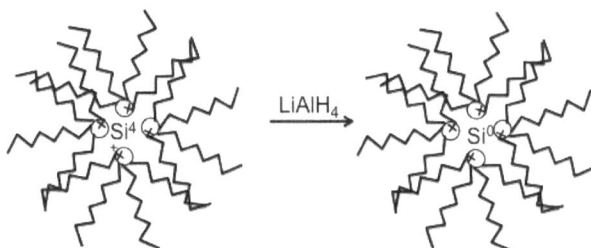

Figure 9. 12: A schematic representation of the inverse micelle used for the synthesis of Si QDs.

After the silicon has been fully reduced and the excess reducing agent quenched, the particles are capped with hydrogens and are hydrophobic. A platinum catalyzed ligand exchange of hydrogen for allylamine will produce hydrophilic particles (Figure 9.13). All reactions in making these particles are extremely air sensitive, and silica is formed readily, so the reactions should be performed in a highly controlled atmosphere, such as a glove box. The particles are then washed in DMF, and finally filtered and stored in deionized water. This will allow the Si QDs to be pure in water, and the particles are ready for analysis. This technique yields Si QDs of 1 - 2 nm in size.

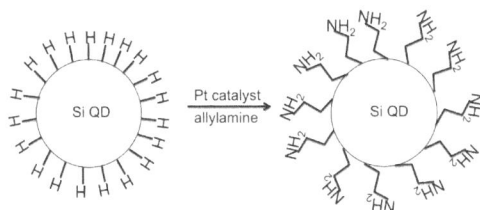

Figure 9.13: Conversion of hydrophobic Si QDs to hydrophillic Si QDs. Adapted from J. H. Warner, A. Hoshino, K. Yamamoto, and R. D. Tilley, Water-soluble photoluminescent silicon quantum dots. *Angew. Chem., Int. Ed.*, 2005, 44, 4550. Copyright: Wiley-VCH (2005).

Sample preparation of Si QDs

The reported absorption wavelength for 1 - 2 nm Si QDs absorb is 300 nm. With the hydrophobic Si QDs, UV-vis absorbance analysis in toluene does not yield an acceptable spectrum because the UV-vis absorbance cutoff is 287 nm, which is very close to 300 nm for the peaks to be resolvable. A better hydrophobic solvent would be hexanes. All measurements of these particles would require a quartz cuvette since the glass absorbance cutoff (300 nm) is exactly where the particles would be observed. Hydrophilic substituted particles do not need to be transferred to another solvent because water's absorbance cutoff is much lower. There is usually a slight impurity of DMF in the water due to residue on the particles after drying. If there is a DMF peak in the spectrum with the Si QDs the wavelengths are far enough apart to be resolved.

What information can be obtained?

Quantum dots are especially interesting when it comes to UV-vis spectroscopy because the size of the quantum dot can be determined from the position of the absorption peak in the UV-vis spectrum. Quantum dots absorb different wavelengths depending on the size of the particles (e.g., Figure 9.14).

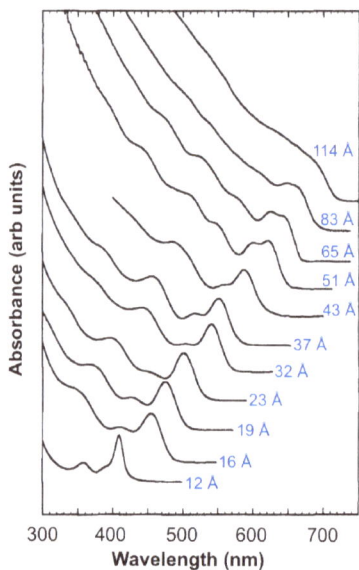

Figure 9.14: A standard absorbance spectrum of different sized CdSe QDs. Adapted from C. B. Murray, D. J. Norris, and M. G. Bawendi, Synthesis and characterization of nearly monodisperse CdE (E = sulfur, selenium, tellurium) semiconductor nanocrystallites. *J. Am. Chem. Soc.*, 1993, 115, 8706. Copyright: American Chemical Society (1993).

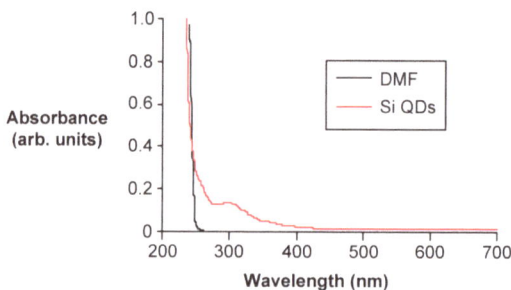

Figure 9.15: UV-visible absorbance spectrum of 1 - 2 nm Si QDs with a DMF reference spectrum.

Many calibration curves would need to be done to determine the exact size and concentration of the quantum dots, but it is entirely possible and very useful to be able to determine size and concentration of quantum dots in this way since other ways of determining size are much more expensive and extensive (electron microscopy is most widely used for this data). An example of silicon quantum dot data can be seen in Figure 9.15. The wider the absorbance peak is, the less monodispersed the sample is.

Why is knowing the size of QDs important?

Different size (different excitation) quantum dots can be used for different applications. The absorbance of the QDs can also reveal how monodispersed the sample is; more monodispersity in a sample is better and more useful in future applications. Silicon quantum dots in particular are currently being researched for making more efficient solar cells. The monodispersity of these quantum dots is particularly important for getting optimal absorbance of photons from the sun or other light source. Different sized quantum dots will absorb light differently, and a more exact energy absorption is important in the efficiency of solar cells. UV-vis absorbance is a quick, easy, and cheap way to determine the monodispersity of the silicon quantum dot sample. The peak width of the absorbance data can give that information. The other important information for future applications is to get an idea about the size of the quantum dots. Different size QDs absorb at different wavelengths; therefore, specific size Si QDs will be required for different cells in tandem solar cells.

Characterization of Group 12-16 quantum dots

Cadmium selenide

Cadmium selenide (CdSe) is one of the most popular Group 12-16 semiconductors for study. This is mainly because the band gap (712 nm or 1.74 eV) energy of CdSe. Thus, the nanoparticles of CdSe can be engineered to have a range of band gaps throughout the visible range, corresponding to the major part of the energy that comes from the solar spectrum. This property of CdSe along with its fluorescing properties is used in a variety of applications such as solar cells and light emitting diodes. Though cadmium and selenium are known carcinogens, the harmful biological effects of CdSe can be overcome by coating the CdSe with a layer of zinc sulfide. Thus CdSe, can also be used as biomarkers, drug-delivery agents, paints and other applications. A typical absorption spectrum of narrow size distribution wurtzite CdSe quantum dot is shown in Figure 9.16.

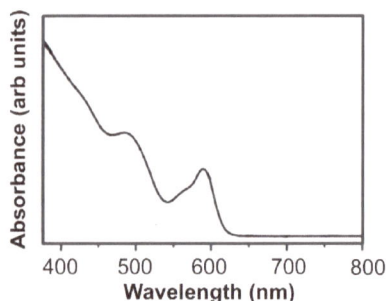

Figure 9.16: UV-visible spectrum of wurtzite CdSe quantum dots. Adapted from X. Zhong, Y. Feng, and Y. Zhang, Facile and reproducible synthesis of red-emitting CdSe nanocrystals in amine with long-term fixation of particle size and size distribution. *J. Phys. Chem. C*, **2007, 111, 526. Copyright: American Chemical Society (2007).**

A series of size evolving absorption spectra are shown in Figure 9.17. However, a complete analysis of the sample is possible only by also studying the fluorescence properties of CdSe.

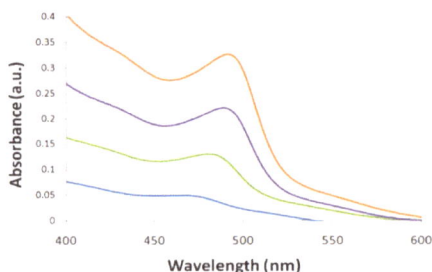

Figure 9.17: Size evolving absorption spectra of CdSe quantum dots.

Cadmium telluride (CdTe)

Cadmium telluride has a band gap of 1.44 eV (860 nm) and as such it absorbs in the infrared region. Like CdSe, the sizes of CdTe can be engineered to have different band edges and thus, different absorption spectra as a function of wavelength. A typical CdTe spectra is shown in Figure 9.18. Due to the small bandgap energy of CdTe, it can be used in tandem with CdSe to absorb in a greater part of the solar spectrum.

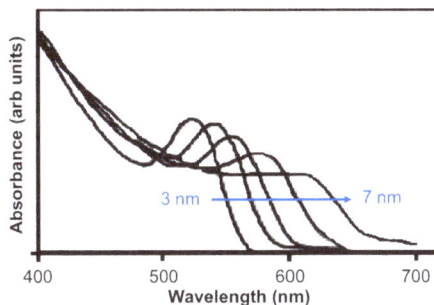

Figure 9.18: Size evolving absorption spectra of CdTe quantum dots from 3 nm to 7 nm. Adapted from Q.-F. Chen, W.-X. Wang, Y.-X. Ge, M.-Y. Li, S.-K. Xu, and X.-J. Zhang, Direct aqueous synthesis of cysteamine-stabilized CdTe quantum dots and its deoxyribonucleic acid bioconjugates. *Chin. J. Anal. Chem.*, 2007, 35, 135. Copyright: Elsevier (2007).

Other Group 12-16 semiconductor systems

Table 9.2 shows the bulk band gap of other Group 12-16 semiconductor systems. The band gap of ZnS falls in the UV region, while those of ZnSe, CdS, and ZnTe fall in the visible region.

Material	Band gap (eV)	Wavelength (nm)
ZnS	3.61	343.2
ZnSe	2.69	460.5
ZnTe	2.39	518.4
CdS	2.49	497.5
CdSe	1.74	712.1
CdTe	1.44	860.3

Table 9.2: Bulk band gaps of different Group 12-16 semiconductors.

Heterostructures of Group 12-16 semiconductor systems

It is often desirable to have a combination of two Group 12-16 semiconductor system quantum heterostructures of different shapes like dots and tetrapods, for applications in solar cells, biomarkers, etc. Some of the most interesting systems are ZnS shell-CdSe core systems, such as the CdSe/CdS rods and tetrapods. Figure 9.19 shows a typical absorption spectrum of CdSe-ZnS core-shell system. This system is important because of the drastically improved fluorescence properties because of the addition of a wide band gap ZnS shell than the core CdSe. In addition, with a ZnS shell, CdSe becomes biocompatible.

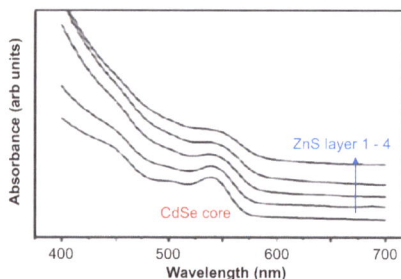

Figure 9.19: Absorption spectra of CdSe core, ZnS shell. Adapted from C.-Q. Zhu, P. Wang, X. Wang and Y. Li, Facile phosphine-free synthesis of CdSe/ZnS core/shell nanocrystals without precursor injection. *Nanoscale Res. Lett.*, 2008, 3, 213. Copyright: Springer (2008).

Characterization of noble metal nanoparticles

Noble metal nanoparticles have been used for centuries to color stained glass windows and provide many opportunities for novel sensing and optical technologies due to their intense scattering (deflection) and absorption of light. One of the most interesting and important properties of noble metal nanoparticles is their localized surface plasmon resonance (LSPR). The LSPR of noble metal nanoparticles arises when photons of a certain frequency induce the collective oscillation of conduction electrons on the nanoparticles' surface. This causes selective photon absorption, efficient scattering, and enhanced electromagnetic field strength around the nanoparticles.

Synthesis of noble metal nanoparticles

Noble metal nanoparticles can be synthesized via the reduction of metal salts. Spherical metal nanoparticle "seeds" are first synthesized by reducing metal salts in water with a strong reducing agent such as sodium borohydride ($NaBH_4$). The seeds are then "capped" to prevent aggregation with a surface group such as citrate (Figure 9.20).

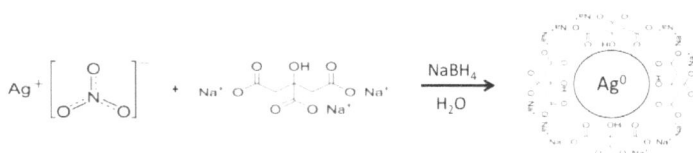

Figure 9.20: Synthesis reaction of citrate-capped silver nanoparticle seeds.

Adjusting the geometry of metal nanoparticles

After small nanoparticle seeds have been synthesized, the seeds can be grown into nanoparticles of various sizes and shapes. Seeds are added to a solution of additional metal salt and a structure-directing agent and are then reduced with a weak reducing agent such as ascorbic acid (see Figure 9.21). The structure-directing agent will determine the geometry of the nanoparticles produced. For example, cetyltrimethylammonium bromide (CTAB) is often used to produce nanorods (Figure 9.21).

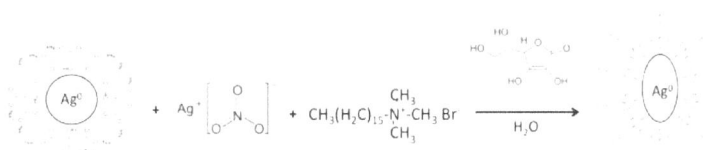

Figure 9.21: Synthesis reaction of cetyltrimethylammonium bromide (CTAB)-capped silver nanorods.

Assemblies of metal nanoparticles

Once synthesized, noble metal nanoparticles can be assembled into various higher-order nanostructures. Nanoparticle dimers, linear chains of two nanoparticles, can be assembled using a linker molecule that binds the two nanoparticles together (Figure 9.22).

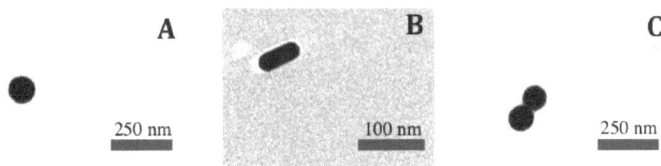

Figure 9.22: TEM images of (a) a gold nanosphere, (b) a gold nanorod and (c) a gold nanosphere dimer.

Less-organized nanoparticle assemblies can be formed through the addition of counterions. Counterions react with the surface groups on nanoparticles, causing the nanoparticles to be stripped of their protective surface coating and inducing their aggregation.

Localized surface plasmon resonance

UV-visible absorbance spectroscopy is a powerful tool for detecting noble metal nanoparticles, because the LSPR of metal nanoparticles allows for highly selective absorption of photons. UV-visible absorbance spectroscopy can also be used to detect various factors that affect the LSPR of noble metal nanoparticles.

Localized surface plasmon resonance (LSPR) is an interesting optical property characteristic of metal nanoparticles, particularly gold nanoparticles (AuNPs), that is exploited for UV-visible spectroscopic characterization. A plasmon is defined as the oscillation of the free electrons in metals. The metal surface is polarized such that positive and negative charges are separated by a distance "d" unique to every metal nanostructure (Figure 9.23).

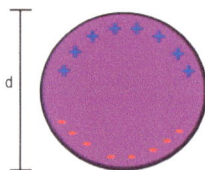

Figure 9.23: A visualization of charge polarization observed in a spherical metal nanoparticle.

If the wavelength of incident radiation matches the separation distance of positive and negative charges, the electrons exhibit resonance. This resonance can be modelled by the mass-spring harmonic oscillator where the negative charges act as a mass on a spring attached to the positive charges (Figure 9.24). The negative charges pull away from the positive charges with some force while the positive charges pull the negative charges back towards them with a restoring force. In larger metal structures, LSPR is localized to one specific part of the structure. However, in metal nanoparticles, such as AuNPs, the LSPR wavelength is often in the UV/visible/IR region (~300-1000 nm).

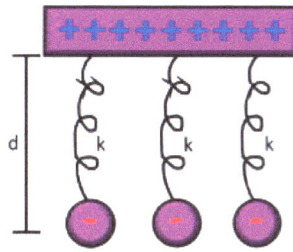

Figure 9.24: Mass spring oscillator model for charge polarization. Resonance occurs when the incident light wavelength (λ) equals the charge separation distance (d), i.e., LSPR occurs at λ = d. The springs have a spring constant (k).

Analyzing AuNPs with UV-visible spectroscopy provides insight into the physical and chemical properties of the sample. When conducting UV-visible spectroscopy, the absorbance is measured over a continuous wavelength range (400 - 800 nm). The peak absorbance occurs at the nanoparticle's specific LSPR wavelength within this range and can shift due to changes in morphology, diameter, surface functionalization, or environment. For instance, spherical AuNPs exhibit LSPR at a wavelength of 525 nm. Since the electrons in AuNPs naturally exhibit SPR at this wavelength, these nanoparticles appear maroon/purple in solution with water or PBS instead of the traditional gold color. Previous studies show that increases in nanoparticle diameter due to either functionalization of the nanoparticle to a polymer or protein or aggregation among individual nanoparticles results in a red shift, a term to indicate that the LSPR wavelength at which peak absorbance is observed increases towards the infrared region. A representative UV-visible spectrum is shown in Figure 9.25.

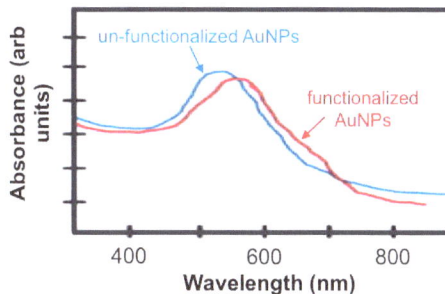

Figure 9.25: UV-visible spectra of un-functionalized AuNPs (blue curve) and functionalized AuNPs (red curve).

Another variable that determines the LSPR wavelength is the morphology of the nanostructure. For instance, the UV-visible spectrogram of gold nanoparticles (AuNP) is different from that of gold nanorods (AuNRs). Bare AuNPs only exhibit one LSPR peak at approximately 525 nm while AuNRs typically exhibit a second LSPR peak wavelength in addition to the first one in the IR region (~800 - 1100 nm), see Figure 9.26. The reason behind the appearance of the second peak is that the electrons can become polarized among the transverse and longitudinal directions of the nanorod. In spherical nanoparticles, the electrons become polarized in only one direction. Figure 9.27 shows the longitudinal and transverse LSPR wavelengths in a sample of gold nanorods. The longitudinal LSPR wavelength is longer than the transverse LSPR wavelength because the distance between the positive and negative charges is longer. This figure represents the charges as both particles and waves to represent the wave-particle duality of electrons. The oscillations of the electric field over time represent how the nanostructure constantly switches polarity and never stays stagnant.

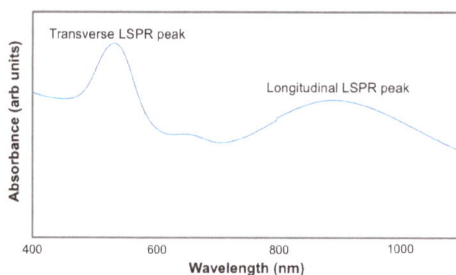

Figure 9.26: A typical UV-visible spectrum for gold nanorods (AuNRs).

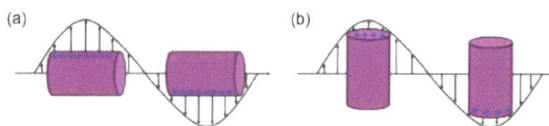

Figure 9.27: Transverse (a) and Longitudinal (b) LSPR for gold nanorods (AuNRs).

When analyzing a sample of metal nanoparticles, UV-visible spectroscopy provides useful information about particle morphology and the success of a functionalization reaction. The LSPR peak wavelength of a sample of non-functionalized metal nanoparticles can be compared to a sample of functionalized nanoparticles to determine whether a red shift occurred. The presence of a red shift could act as an indicator of successful functionalization. However, one of the limitations of UV-visible spectroscopy is that it cannot accurately quantify the extent of metal nanoparticle functionalization. The LSPR peak wavelength can be shifted due to changes in the environment that the metal nanoparticles exist in along with changes in morphology. It can be difficult to distinguish which factor dominates the UV-visible peak shift because it is possible for unconjugated polymer or protein, reaction intermediates, reactants, or other impurities to float around in solution along with the conjugated nanoparticles. Conjugation reactions rarely yield 100% functionalized metal nanoparticles.

Finally, UV-visible spectroscopy cannot unambiguously account for nanoparticle aggregation, which can occur due to interactions between the polymer or protein conjugated to the nanoparticle surface or the ion/ion interactions between the nanoparticles themselves. There are numerous factors that can account for an LSPR peak shift in UV-visible, so it is not possible to attribute the shift to aggregation without further analysis. Fortunately, UV-visible is often conducted in conjunction with DLS to determine the presence of nanoparticle aggregation and the approximate size of the spherical nanoparticle.

Mie theory

Mie theory, a theory that describes the interaction of light with a homogenous sphere, can be used to predict the UV-visible absorbance spectrum of spherical metallic nanoparticles. One equation that can be obtained using Mie theory which describes the extinction, the sum of absorption and scattering of light, of spherical nanoparticles,

$$E(\lambda) = \frac{24\pi N_A a^3 \varepsilon_m^{3/2}}{\lambda \ln(10)} \left[\frac{\varepsilon_i}{(\varepsilon_r + 2\varepsilon_m)^2 + \varepsilon_i^2} \right]$$

where, $E(\lambda)$ is the extinction, N_A is the areal density of the nanoparticles, a is the radius of the nanoparticles, ε_m is the dielectric constant of the environment surrounding the nanoparticles, λ is the wavelength of the incident light, and ε_r and ε_i are the real and imaginary parts of the nanoparticles' dielectric function. From this relation, we can see that the UV-visible absorbance spectrum of a solution of nanoparticles is dependent on the radius of the nanoparticles, the composition of the nanoparticles, and the environment surrounding the nanoparticles.

More advanced theoretical techniques

Mie theory is limited to spherical nanoparticles, but there are other theoretical techniques that can be used to predict the UV-visible spectrum of more complex noble metal nanostructures. These techniques include surface-based methods such as the generalized multipole technique and T-matrix method, as well as volume-based techniques such as the discrete dipole approximation and the finite different time domain method.

Prediction of geometry

Just as the theoretical techniques described above can use nanoparticle geometry to predict the UV-visible absorbance spectrum of noble metal nanoparticles, nanoparticles' UV-visible absorbance spectrum can be used to predict their geometry. As shown in Figure 9.28, the UV-visible absorbance spectrum is highly dependent on nanoparticle geometry.

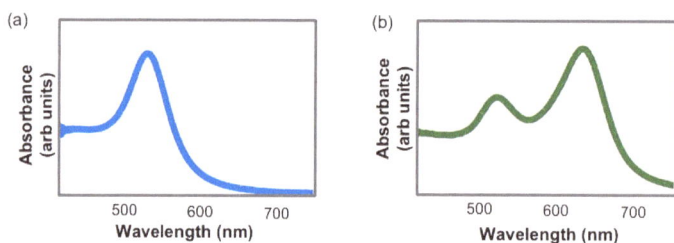

Figure 9.28: UV-visible absorbance spectra of (a) 50 nm diameter gold nanospheres and (b) gold nanorods of 60 nm × 25 nm (length × diameter).

The shapes of the two spectra are quite different despite the two types of nanoparticles having similar dimensions and being composed of the same material (Figure 9.23).

Determination of aggregation states

The UV-visible absorbance spectrum is also dependent on the aggregation state of the nanoparticles. When nanoparticles are in close proximity to each other, their plasmons couple, which affects their LSPR and thus their absorption of light. Dimerization of nanospheres causes a "red shift," a shift to longer wavelengths, in the UV-visible absorbance spectrum as well as a slight increase in absorption at higher wavelengths (see Figure 9.29). Unlike dimerization, aggregation of nanoparticles causes a decrease in the intensity of the peak absorbance without shifting the wavelength at which the peak occurs (λ_{max}). Figure 9.29 illustrates the increase in nanoparticle aggregation with increased salt concentrations based on the decreased absorbance peak intensity.

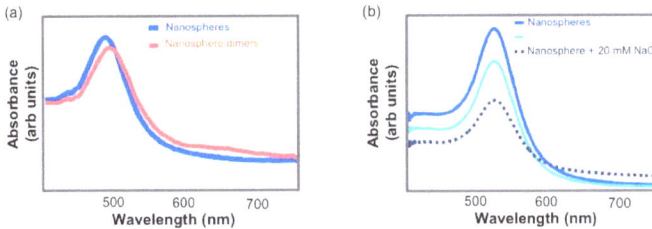

Figure 9.29: UV-visible absorbance spectrum of (a) 50 nm gold nanosphere dimers with a reference spectrum of single gold nanospheres and (b) 50 nm gold nanospheres exposed to various concentrations of NaCl.

Determination of surface composition

The λ_{max} of the UV-visible absorbance spectrum of noble metal nanoparticles is highly dependent on the environment surrounding the nanoparticles. Because of this, shifts in λ_{max} can be used to detect changes in the surface composition of the nanoparticles. One is using UV-visible absorbance spectroscopy to detect the binding of biomolecules to the surface of noble metal nanoparticles. The red shift in the λ_{max} of the UV-visible absorbance spectrum in Figure 9.30 with the addition of human serum albumin protein indicates that the protein is binding to the surface of the nanoparticles.

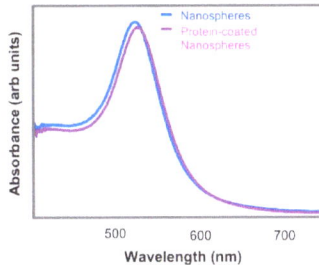

Figure 9.30: UV-visible absorbance spectrum of 50 nm gold nanospheres exposed to human serum albumin protein with a reference spectrum of nanospheres exposed to deionized water.

Optical characterization by fluorescence spectroscopy

Emission spectroscopy, in general, refers to a characterization technique that measures the emission of radiation by a material that has been excited. Fluorescence spectroscopy is one type of emission spectroscopy which records the intensity of light radiated from the material as a function of wavelength. It is a nondestructive characterization technique.

After an electron is excited from the ground state, it needs to relax back to the ground state. This relaxation or loss of energy to return to the ground state, can be achieved by a combination of non-radiative decay (loss of energy through heat) and radiative decay (loss of energy through light). Non-radiative decay by vibrational modes typically occurs between energy levels that are close to each other. Radiative decay by the emission of light occurs when the energy levels are far apart like in the case of the band gap. This is because loss of energy through vibrational modes across the band gap can result in breaking the bonds of the crystal. This phenomenon is shown in Figure 9.31.

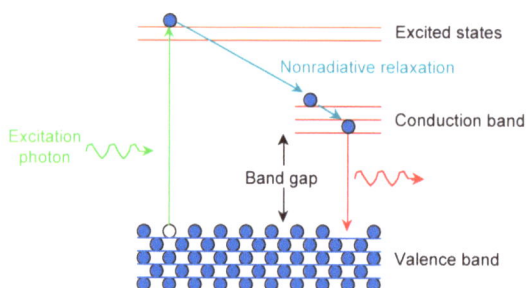

Figure 9.31: Emission of luminescence photon for Group 12-16 semiconductor quantum dot.

The band gap of Group 12-16 semiconductors is in the UV-visible region. Thus, the wavelength of the emitted light as a result of radiative decay is also in the visible region, resulting in fascinating fluorescence properties.

A fluorimeter is a device that records the fluorescence intensity as a function of wavelength. The fluorescence quantum yield can then be calculated by the ratio of photons absorbed to photons emitted by the system. The quantum yield gives the probability of the excited state getting relaxed via fluorescence rather than by any other non-radiative decay.

Difference between fluorescence and phosphorescence

Photoluminescence is the emission of light from any material due to the loss of energy from excited state to ground state. There are two main types of luminescence – fluorescence and phosphorescence. Fluorescence is a fast decay process, where the emission rate is around $10^8\,s^{-1}$ and the lifetime is around 10^{-9} - 10^{-7} s. Fluorescence occurs when the excited state electron has an opposite spin compared to the ground state electrons. From the laws of quantum mechanics, this is an allowed transition, and occurs rapidly by emission of a photon. Fluorescence disappears as soon as the exciting light source is removed.

Phosphorescence is the emission of light, in which the excited state electron has the same spin orientation as the ground state electron. This transition is a forbidden one and hence the emission rates are slow (10^3 - $10^0\,s^{-1}$). So, the phosphorescence lifetimes are longer, typically seconds to several minutes, while the excited phosphors slowly returned to the ground state. Phosphorescence is still seen, even after the exciting light source is removed. Group 12-16 semiconductor quantum dots exhibit fluorescence properties when excited with ultraviolet light.

Instrumentation

The working schematic for the fluorometer is shown in Figure 9.32.

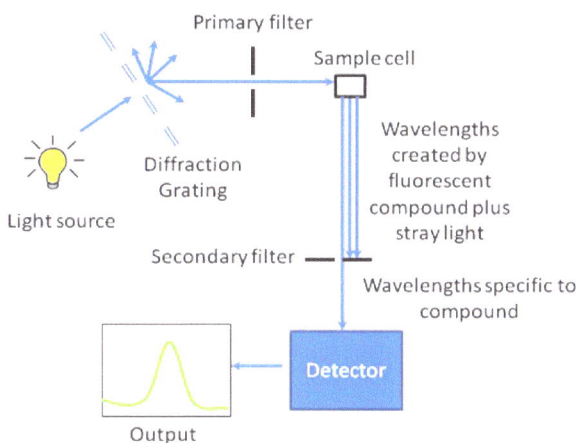

Figure 9.32: Schematic of fluorometer.

The light source

The excitation energy is provided by a light source that can emit wavelengths of light over the ultraviolet and the visible range. Different light sources can be used as excitation sources such as lasers, xenon arcs and mercury-vapor lamps. The choice of the light source depends on the sample. A laser source emits light of a high irradiance at a very narrow wavelength interval. This makes the need for the filter unnecessary, but the wavelength of the laser cannot be altered significantly. The mercury vapor lamp is a discrete line source. The xenon arc has a continuous emission spectrum between the ranges of 300 - 800 nm.

The diffraction grating and primary filter

The diffraction grating splits the incoming light source into its component wavelengths (Figure 9.32). The monochromator can then be adjusted to choose with wavelengths to pass through. Following the primary filter, specific wavelengths of light are irradiated onto the sample

Secondary filter

The secondary filter is placed at a 90° angle (Figure 9.32) to the original light path to minimize the risk of transmitted or reflected incident light reaching the detector. Also, this minimizes the amount of stray light, and results in a better signal-to-noise ratio. From the secondary filter, wavelengths specific to the sample are passed onto the detector.

Detector

The detector can either be single-channeled or multichanneled (Figure 9.32). The single-channeled detector can only detect the intensity of one wavelength at a time, while the multichanneled detects the intensity at all wavelengths simultaneously, making the emission monochromator or filter unnecessary. The different types of detectors have both advantages and disadvantages.

Output

The output is the form of a plot of intensity of emitted light as a function of wavelength as shown in Figure 9.33.

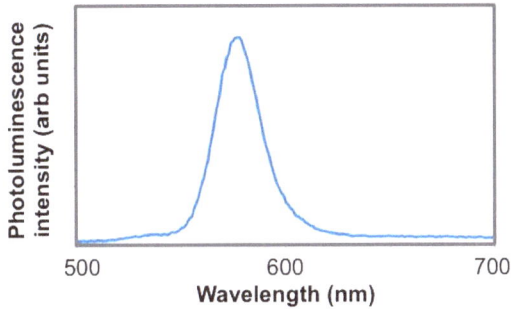

Figure 9.33: Emission spectra of CdSe quantum dot.

Analysis of data

The data obtained from fluorimeter is a plot of fluorescence intensity as a function of wavelength. Quantitative and qualitative data can be obtained by analysing this information.

Quantitative information

From the fluorescence intensity versus wavelength data, the quantum yield (Φ_F) of the sample can be determined. Quantum yield is a measure of the ratio of the photons absorbed with respect to the photons emitted. It is important for the application of Group 12-16 semiconductor quantum dots using their fluorescence properties, for e.g., biomarkers.

The most well-known method for recording quantum yield is the comparative method which involves the use of well characterized standard solutions. If a test sample and a standard sample have similar absorbance values at the same excitation wavelength, it can be assumed that the number of photons being absorbed by both the samples is the same. This means that a ratio of the integrated fluorescence intensities of the test and standard sample measured at the same excitation wavelength will give a ratio of quantum yields. Since the quantum yield of the standard solution is known, the quantum yield for the unknown sample can be calculated.

A plot of integrated fluorescence intensity versus absorbance at the excitation wavelength is shown in Figure 9.34. The slope of the graphs shown in Figure 9.34 are proportional to the quantum yield of the different samples. Quantum yield is then calculated using,

$$\frac{QY_X}{QY_{ST}} = \frac{slope_X}{slope_{ST}} \frac{(RI_X)^2}{(RI_{ST})^2}$$

where subscripts ST denotes standard sample and X denotes the test sample; QY is the quantum yield; RI is the refractive index of the solvent.

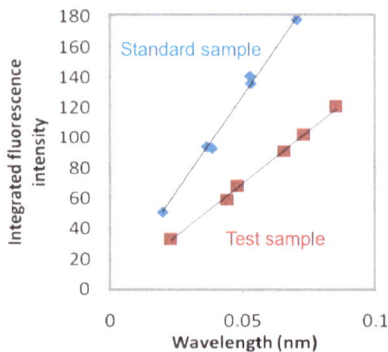

Figure 9.34: Integrated fluorescence intensity as a function of absorbance.

Take the example of Figure 9.34, if the same solvent is used in both the sample and the standard solution, the ratio of quantum yields of the sample to the standard is given by,

$$\frac{QY_X}{QY_{ST}} = \frac{1.41}{2.56}$$

If the quantum yield of the standard is known to 0.95, then the quantum yield of the test sample is 0.523 or 52.3%.

The assumption used in the comparative method is valid only in the Beer-Lambert law linear regime. Beer-Lambert law states that absorbance is directly proportional to the path length of light travelled within the sample, and concentration of the sample. The factors that affect the quantum yield measurements are the following:

- **Concentration** – Low concentrations should be used (absorbance < 0.2 a.u.) to avoid effects such as self-quenching.
- **Solvent** – It is important to take into account the solvents used for the test and standard solutions. If the solvents used for both are the same, then the comparison is trivial. However, if the solvents in the test and

standard solutions are different, this difference needs to be accounted for. This is done by incorporating the solvent refractive indices in the ratio calculation.

- **Standard samples:** The standard samples should be characterized thoroughly. In addition, the standard sample used should absorb at the excitation wavelength of the test sample.
- **Sample preparation:** It is important that the cuvettes used are clean, scratch free and clear on all four sides. The solvents used must be of spectroscopic grade and should not absorb in the wavelength range.
- **Slit width:** The slit widths for all measurements must be kept constant.

The quantum yield of the Group 12-16 semiconductor nanoparticles is affected by many factors such as the following.

- **Surface defects:** The surface defects of semiconductor quantum dots occur in the form of unsatisfied valencies. Thus, resulting in unwanted recombination. These unwanted recombinations reduce the loss of energy through radiative decay, and thus reducing the fluorescence.
- **Surface ligands:** If the surface ligand coverage is a 100%, there is a smaller chance of surface recombination to occur.
- **Solvent polarity:** If the solvent and the ligand have similar solvent polarities, the nanoparticles are more dispersed, reducing the loss of electrons through recombination.

Qualitative Information

Apart from quantum yield information, the relationship between intensity of fluorescence emission and wavelength, other useful qualitative information such as size distribution, shape of the particle and presence of surface defects can be obtained.

As shown in Figure 9.35, the shape of the plot of intensity versus wavelength is a Gaussian distribution. In Figure 9.35 the full width at half maximum (FWHM) is given by the difference between the two extreme values of the wavelength at which the photoluminescence intensity is equal to half its maximum value. From the full width half max (FWHM) of the fluorescence intensity Gaussian distribution, it is possible to determine qualitatively the size distribution of the sample. For a Group 12-16 quantum dot sample if the FWHM is greater than 30, the system is very polydisperse and has a large size distribution. It is desirable for all practical applications for the FWHM to be lesser than 30.

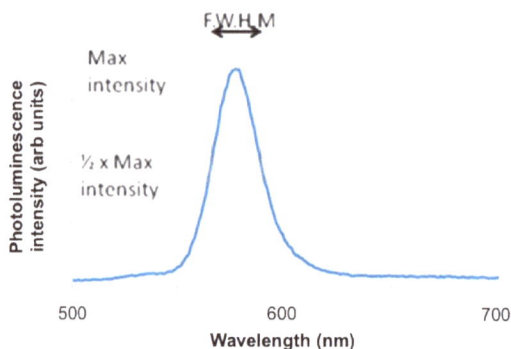

Figure 9.35: Emission spectra of CdSe QDs showing the full width half maximum (FWHM).

From the FWHM of the emission spectra, it is also possible to qualitatively get an idea if the particles are spherical or shaped. During the synthesis of the shaped particles, the thickness of the rod or the arm of the tetrapod does not vary among the different particles, as much as the length of the rods or arms changes. The thickness of the arm or rod is responsible for the quantum effects in shaped particles. In the case of quantum dots, the particle is quantum confined in all dimensions. Thus, any size distribution during the synthesis of quantum dots greatly affects the emission spectra. As a result, the FWHM of rods and tetrapods are much smaller as compared to a quantum dot. Hence, qualitatively it is possible to differentiate between quantum dots and other shaped particles.

Another indication of branched structures is the decrease in the intensity of fluorescence peaks. Quantum dots have very high fluorescence values as compared to branched particles, since they are quantum confined in all dimensions as compared to just 1 or 2 dimensions in the case of branched particles.

Group 12-16 semiconductor nanoparticles

The emission spectra of all Group 12-16 semiconductor nanoparticles are Gaussian curves as shown in Figure 9.35. The only difference between them is the band gap energy, and hence each of the Group 12-16 semiconductor nanoparticles fluoresce over different ranges of wavelengths

Cadmium selenide

Since its bulk band gap (1.74 eV, 712 nm) falls in the visible region cadmium selenide (CdSe) is used in various applications such as solar cells, light emitting diodes, etc. Size evolving emission spectra of cadmium selenide is shown in Figure 9.36. Different sized CdSe particles have different colored fluorescence spectra. Since cadmium and selenide are known carcinogens and being nanoparticles are easily absorbed into the human body, there is some concern regarding these particles.

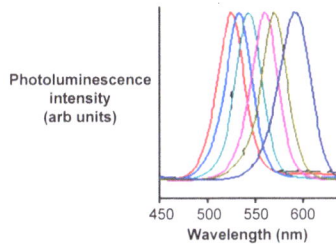

Figure 9.36: Size evolving CdSe emission spectra. Adapted from http://www.physics.mq.edu.au.

A combination of the absorbance and emission spectra is shown in Figure 9.37 for four different sized particles emitting green, yellow, orange, and red fluorescence.

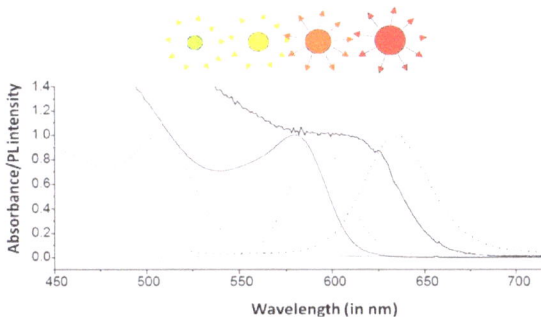

Figure 9.37: Absorption and emission spectra of CdSe quantum dots. Adapted from G. Schmid, *Nanoparticles: From Theory to Application*, Wiley-VCH, Weinham (2004). Copyright: Wiley-VCH (2004).

Cadmium telluride

Cadmium telluride (CdTe) has a band gap of 1.44 eV and thus absorbs in the infra-red region. The size evolving CdTe emission spectra is shown in Figure 9.38.

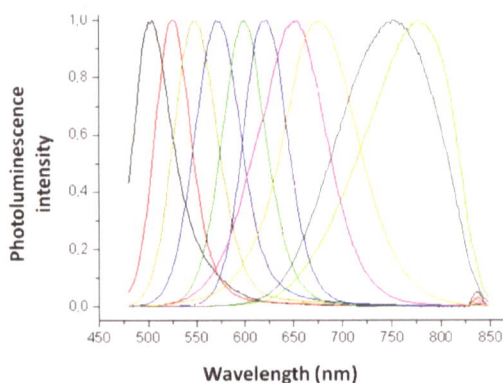

Figure 9.38: Size evolution spectra of CdTe quantum dots.

Adding shells to QDs

Capping a core quantum dot with a semiconductor material with a wider bandgap than the core, reduces the nonradiative recombination and results in brighter fluorescence emission. Quantum yields are affected by the presences of free surface charges, surface defects and crystal defects, which results in unwanted recombination. The addition of a shell reduces the nonradiative transitions and majority of the electrons relax radiatively to the valence band. In addition, the shell also overcomes some of the surface defects.

For the CdSe-core/ZnS-shell systems exhibit much higher quantum yield as compared to core CdSe quantum dots as seen in Figure 9.39.

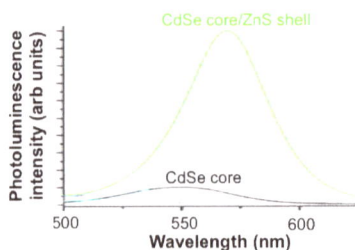

Figure 9.39: Emission spectra of core CdSe only and CdSe-core/ZnS-shell.

Bibliography

V. Amendola, R. Pilot, M. Frasconi, O. M. Marago, and M. A. Iati, Surface plasmon resonance in gold nanoparticles: a review. *J. Phys. Condens. Matter*, 2017, **29**, 203002.

X. Blase, A. Rubio, S. G. Louie, and M. L. Cohen, Quasiparticle band structure of bulk hexagonal boron nitride and related systems. *Phys. Rev. B*, 1995, **51**, 6868.

D. K. Bozanic, A. S. Luyt, L. V. Trandafilovic, and V. Djokovic, Glycogen and gold nanoparticlebioconjugates: controlled plasmon resonance *via*glycogen-induced nanoparticle aggregation. *RSC Adv.*, 2013, **3**, 8705.

Y. Canivez, Quick and easy measurement of the band gap in semiconductors. *Eur. J. Phys.* 1983, **4**, 42.

J. Cao, T. Sun, and K. T. V. Grattan, Gold nanorod-based localized surface plasmon resonance biosensors: a review. *Sens. Actuators B Chem.*, 2014, **195**, 332.

Q.-F. Chen, W.-X. Wang, Y.-X. Ge, M.-Y. Li, S.-K. Xu, and X.-J. Zhang, Direct aqueous synthesis of cysteamine-stabilized CdTe quantum dots and its deoxyribonucleic acid bioconjugates. *Chin. J. Anal. Chem.*, 2007, **35**, 135.

B. O. Dabbousi, J. Rodriguez-Viejo, F. V. Mikulec, J. R. Heine, H. Mattoussi, R. Ober, K. F. Jensen, and M. G. Bawendi, (CdSe)ZnS core–shell quantum dots: synthesis and characterization of a size series of highly luminescent nanocrystallites. *J. Phys. Chem. B*, 1997, **101**, 9463.

S. V. Gapoenko, *Optical Properties of Semiconductor Nanocrystals*, Cambridge University Press, Cambridge (2003).

J. Haes, S. Zou, G. C. Schatz, and R. P. Van Duyne, A nanoscale optical biosensor: the long range distance dependence of the localized surface plasmon resonance of noble metal nanoparticles. *J. Phys. Chem. B*, 2004, **108**, 109.

W. Haiss, N. T. K. Tanh, J. Aveyard, and D. G. Fernig, Determination of size and concentration of gold nanoparticles from UV–vis spectra. *Anal. Chem.*, 2007, **79**, 4215.

J. Y. Hariba, *A Guide to Recording Fluorescence Quantum Yield*, Jobin Yvon Hariba Limited, Stanmore (2003).

C. J. Murphy, T. K. Sau, A. M. Gole, C. J. Orendorff, J. Gao, L. Gou, S. E. Hunyadi, and T. Li, Anisotropic metal nanoparticles: synthesis, assembly, and optical applications. *J. Phys. Chem. B*, 2005, **109**, 13857.

C. B. Murray, D. J. Norris, and M. G. Bawendi, Synthesis and characterization of nearly monodisperse CdE (E = sulfur, selenium, tellurium) semiconductor nanocrystallites. *J. Am. Chem. Soc.*, 1993, **115**, 8706.

B. L. Oliva-Chatelain and A. R. Barron, Experiments towards size and dopant control of germanium quantum dots for solar applications. *AIMS Mater. Sci.*, 2016, **3**, 1

B. L. Oliva-Chatelain, T. M. Ticich, and A. R. Barron, Doping silicon nanocrystals and quantum dots. *Nanoscale*, 2016, **8**, 1733.

A. W. Orbaek, M. McHale, and A. R. Barron, Synthesis and characterization of silver nanoparticles for an undergraduate laboratory. *J. Chem. Edu.*, 2015, **92**, 339.

A. L. Rogach, *Semiconductor Nanocrystal Quantum Dots. Synthesis, Assembly, Spectroscopy and Applications*, Springer Wien, New York (2008).

T. Pradeep, *Nano: The Essentials. Understanding Nanoscience and Nanotechnology*, Tata McGraw-Hill, New Delhi (2007).

G. Raschke, S. Kowarik, T. Franzl, C. Sönnichsen, T. A. Klar, and J. Feldmann, Biomolecular recognition based on single gold nanoparticle light scattering. *Nano Lett.*, 2003, **3**, 935.

R. A. Reynolds, C. A. Mirkin, R. A. Letsinger, Homogeneous, nanoparticle-based quantitative colorimetric detection of oligonucleotides. *J. Am. Chem. Soc.,* 2000, **122**, 3795.

P. Sarkar, D. K. Bhui, H. Bar, G. P. Sahoo, S. Samanta, S. Pyne, and A. Misra, Aqueous-phase synthesis of silver nanodiscs and nanorods in methyl cellulose matrix: photophysical study and simulation of UV–vis extinction spectra using DDA method. *Nanoscale Res. Lett.,* 2010, **5**, 161.

T. Sato, Y. Yamamoto, Y. Fujishiro, and S. Uchida, Intercalation of iron oxide in layered $H_2Ti_4O_9$ and $H_4Nb_6O_{17}$: visible-light induced photocatalytic properties. *J. Chem. Soc., Faraday Trans.,* 1996, **92**, 5089.

D. A. Stuart, A. J. Hayes, C. R. Yonzon, E. M. Hicks, and R. P. Van Duyne, Biological applications of localised surface plasmonic phenomenae. *IEE Proc.-Nanobiotechnol.,* 2005, **152**, 13.

T. Sato, K. Masaki, K. Sato, Y. Fujishiro, and A. Okuwaki, Photocatalytic properties of layered hydrous titanium oxide/CdS-ZnS nanocomposites incorporating CdS-ZnS into the interlayer. *J. Chem. Tech. Biotechnol.,* 1996, **67**, 339.

G. Schmid, *Nanoparticles: From Theory to Application*, Wiley-VCH, Weinheim (2004).

D. V. Talapin, J. H. Nelson, E. V. Shevchenko, S. Aloni, B. Sadtler, and A. P. Alivisatos, Seeded growth of highly luminescent CdSe/CdS nanoheterostructures with rod and tetrapod morphologies. *Nano Lett.*, 2007, 7, 2951.

J. Tauc, R. Grigorovici, and A. Vancu, Optical properties and electronic structure of amorphous germanium. *Phys. Status Solidi.* 1996, **15**, 627.

J. H. Warner, A. Hoshino, K. Yamamoto, and R. D. Tilley, Water-soluble photolu-minescent silicon quantum dots. *Angew. Chem., Int. Ed.*, 2005, **44**, 4550.

A. T. R. Williams, S. A. Winfield, and J. N. Miller, Relative fluorescence quantum yields using a computer-controlled luminescence spectrometer. *Analyst*, 1983, **108**, 1067.

W. W. Yu, L. Qu, W. Guo, and X. Peng, Experimental determination of the extinction coefficient of CdTe, CdSe, and CdS nanocrystals. *Chem. Mater.*, 2003, **15**, 2854.

X. Zhong, Y. Feng, and Y. Zhang, Facile and reproducible synthesis of red-emitting CdSe nanocrystals in amine with long-term fixation of particle size and size dis-tribution. *J. Phys. Chem. C*, 2007, **111**, 526.

C.-Q. Zhu, P. Wang, X. Wang and Y. Li, Facile phosphine-free synthesis of CdSe/ZnS core/shell nanocrystals without precursor injection. *Nanoscale Res. Lett.*, 2008, **3**, 213.

Chapter 10: Dynamic Light Scattering

Angel Adrian Garces, Pavan M. V. Raja and Andrew R. Barron

Principles of dynamic light scattering (DLS)

Dynamic light scattering (DLS) provides information about nanoparticle size, as determined by the hydrodynamic diameter, and aggregation, as determined by the polydispersity index. Additionally, DLS describes the different nanoparticle sizes in solution through the creation of an intensity weighted gaussian distribution. These quantities are obtained by measuring the changes in the intensity of scattered monochromatic laser light by the sample at a fixed incident angle. The scattered light passes through a detector at a certain angle to the incident light. For example, DynaPro® NanoStar® instruments place a detector at 90° while Zetasizer Nano S® from Malvern Instruments places a backscatter detection system at 173° to the incident light. The advantage of a backscatter system is that it allows for the analysis of samples at higher concentrations because it avoids the multiple scattering phenomenon, which describes the "scattering of a photon by multiple particles." A sample is analyzed in a clear cuvette so that the laser light runs through the sample and scatters at varying intensities. The variation in scattered light intensity occurs because the particles in the sample are continuously moving in the sample according to the principles of Brownian motion, a theory that describes the random directional flow of nanoparticles in solution. Figure 10.1 shows the Brownian motion of particles within the fixed boundaries of a cube.

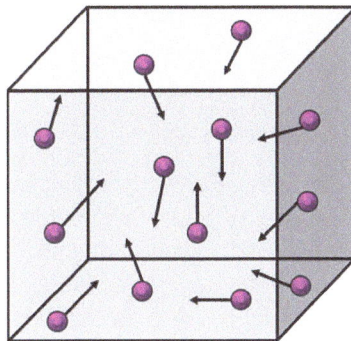

Figure 10.1: A schematic of Brownian motion of metal nanoparticles in solution.

Hydrodynamic diameter and polydispersity index

The hydrodynamic diameter of a spherical nanoparticle (NP) is defined as the diameter of the solvated particle as it moves around in solution according to the principles of Brownian motion. The size of the solvation shell, and hence the error in NP diameter reported by DLS, varies depending on the solvent. To minimize this error, NPs are commonly suspended in a solution of PBS or a solution of NaCl and water for DLS analysis. Subjecting functionalized NPs to DLS analysis and comparing the hydrodynamic diameter to that of bare NPs can provide insight as to whether functionalization was successful. For example, studies showed that the hydrodynamic diameter of a bare 60 nm AuNP colloid compared to AuNPs conjugated with IgG antibody increased by 16 nm. The hydrodynamic diameter of a conjugated nanoparticle can be predicted by adding the size of the species to the diameter of the bare nanoparticle. However, this assumes that an even, single layer coating on the nanoparticle. DLS must be conducted to determine the true hydrodynamic diameter and extent of functionalization.

The polydispersity index is a measure how much variation in AuNP diameter exists in solution. Using the intensity weighted Gaussian distribution, the polydispersity index is defined as the ratio of the standard deviation to the mean of the sample. This value ranges from 0 - 1.0. Lower values (0.05 - 0.10) describe uniform solutions with little variety in nanoparticle diameter while higher values (>0.4) indicates that the solution has a wide variety of nanoparticle diameters. While the polydispersity index quantifies the dispersity present in solution, the intensity correlation function and intensity weighted Gaussian distribution creates a visual representation of the particle size.

Interpreting the intensity correlation function and volume-weighted Gaussian distribution

The intensity correlation function measures signal intensity, which varies due to the continuous movement of particles in solution, over time at a given incident angle of light. Particles with smaller diameters scatter light at a higher rate than particles with larger diameters. As a result, smaller particles yield a graph characterized with sharper peaks than larger particles (Figure 10.2). When determining the success of nanoparticle functionalization, the intensity correlation function should become less jagged for functionalized nanoparticles compared to bare nanoparticles due to the increase in diameter. The disadvantage of relying solely on this function is that it does not provide any information about particle aggregation. To obtain information on aggregation, the intensity correlation function is converted to the intensity weighted Gaussian distribution through a set of mathematical transformations that will not be discussed at depth in this module. Finally, other variables to consider when optimizing data collection for NP size analysis are the wavelength of monochromatic laser light used, the incident angle of light, the temperature, concentration, and viscosity of the solution, and the presence of dust in and around the sample and light source.

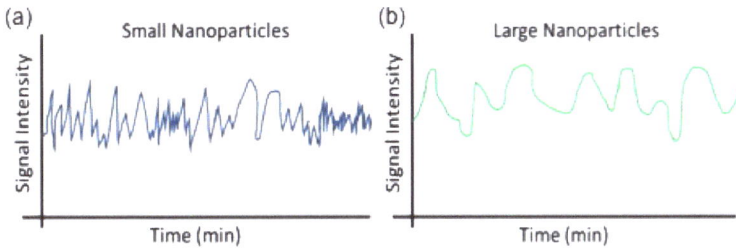

Figure 10.2: The intensity correlation function changes with NP diameter. Smaller NPs produced a jagged plot (a) and larger NPs produced a smoother plot (b).

The intensity weighted Gaussian distribution measures volume distribution, expressed as a percentage, against diameter, expressed in nanometers. The volume distribution is a distribution of the percent volume that specific diameters of nanoparticles occupy in the solution. The Gaussian distribution provides a size specific distribution of nanoparticles and helps to determine the presence of nanoparticle aggregation. Unfortunately, it does not provide a quantitative measure of particle aggregation, solely whether it exists or not. It is up to the researcher to determine the reason for aggregation. While there are several different mechanisms of action that could cause nanoparticle aggregation, there are some common themes present. Protein-protein interactions between different hydrophobic or hydrophilic amino acids conjugated to nanoparticles could result in the formation of nanoparticle clusters. This is especially important to consider in drug delivery applications of nanoparticles because the hydrodynamic diameter affects whether the drug will reach the intended target or not. According to studies, a hydrodynamic diameter of less than 200 nm and a monodisperse solution are ideal for intravenous injection. Other potential sources of particle aggregation include ion-ion interactions between individual nanoparticles themselves or aggregation between individual nanoparticles mediated by solvent molecules.

Particle aggregation can be detected in the Gaussian distribution by the appearance of multiple peaks (Figure 10.3). A uniform solution exhibits one peak centered around a mean diameter while a solution with aggregates often exhibits multiple peaks with multiple mean diameters. A two peak graph with one peak centered at ~50-80 nm and another peak centered at ~120-200 nm is common for nanoparticles that are prone to aggregation. The addition of a detergent to the nanoparticle solution, such as Tween-20 (Figure 10.4) or SDS (Figure 10.5), prevents the nanoparticles from interacting with each other and hence increases the solution homogeneity while decreasing the polydispersity index. While particle aggregation certainly plays a role in the data quality of the volume-weighted distribution, concentration is another important factor to consider. An overly dilute solution will not scatter incident monochromatic light strongly enough to create a clean intensity correlation function, which will distort the volume-weighted distribution. An overly concentrated solution means that particles exist in closer proximity to each other. As a result, the particle motion strays away from Brownian motion, which is a core assumption for DLS to provide accurate data.

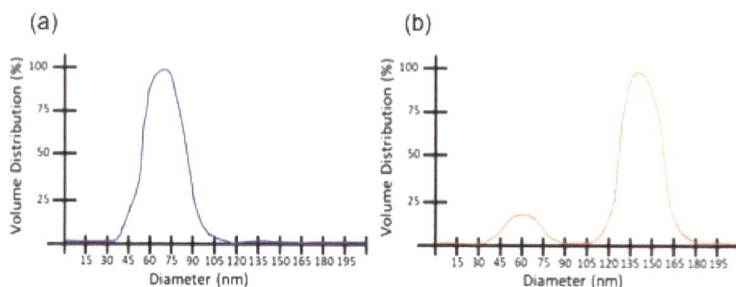

Figure 10.3: The shape of the Volume-Weighted Gaussian Distribution depends on the extent of nanoparticle aggregation. A monodisperse solution with minimal aggregation displays a single peak (a), while an aggregated solution displays multiple peaks (b).

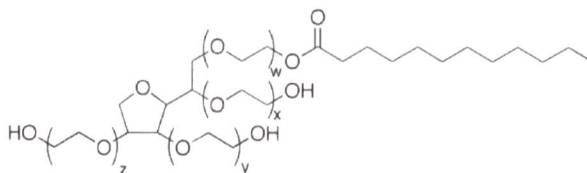

Figure 10.4: The structure of polyoxyethylene (20) sorbitan monolaurate (Tween 20), where $w + x + y + z = 20$.

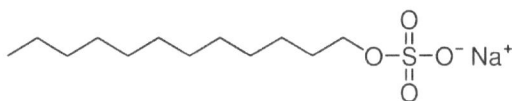

Figure 10.5: The structure of sodium dodecyl sulfate (SDS).

Advantages and disadvantages of DLS

DLS has numerous advantages over other nanoparticle visualization techniques. First, DLS is conducted in solution and requires only ~20 uL of sample in ~5 mL solution to provide an accurate measurement of the hydrodynamic diameter and polydispersity index. In addition, the technique is non-destructive to nanoparticle samples since only monochromatic light is passed through it. The same sample can be reanalyzed several times if needed. Other nanoparticle visualization techniques, such as scanning electron microscopy (SEM), often employ freeze-drying and contrast agents when visualizing nanoparticle interaction with cells. These agents can significantly affect the nanoparticle size. Finally, DLS is relatively inexpensive and easy to run compared to other analytical techniques for nanoparticles. As a result, it has become one of the most commonly used analytical techniques for nanoparticle sizing.

While DLS is a powerful technique that provides accurate information about nanoparticle size and aggregation in the form of a Gaussian distribution, it does not come without its disadvantages. One downside of DLS is that it does not allow the user to directly visualize the morphology of the nanoparticles. DLS provides information about the existence of nanoparticle aggregates in solution via the volume-weighted Gaussian distribution. However, the user cannot determine the shape of the aggregates without the use of a microscopy technique, such as transmission electron microscopy (TEM) or confocal microscopy. Morphology is a crucial aspect to understand in determining the type of interactions that are causing nanoparticle aggregation. Another disadvantage to DLS for analyzing nanoparticle size is low peak resolution. To accurately separate two distinct nanoparticle sizes, they must differ by at least 50% for a separate peak to appear in the volume-weighted distribution. As a result, some nanoparticles in solution might not be represented in the final distribution. Finally, the sample cuvette and laser must be devoid of dust particles as these will appear in the volume-weighted Gaussian distribution if the cuvette and laser are not cleaned carefully.

Bibliography

B. J. Berne and R. Pecora, *Dynamic Light Scattering: With Applications to Chemistry, Biology, and Physics.* Courier Corporation, 2000.

A. Bootz, V. Vogel, D. Schubert, and J. Kreuter, Comparison of scanning electron microscopy, dynamic light scattering and analytical ultracentrifugation for the sizing of poly(butyl cyanoacrylate) nanoparticles. *Eur. J. Pharm. Biopharm.*, 2004, **57**, 369.

CPS Instruments Europe. Comparison of Particle Sizing Methods. http://www.cpsin-struments.eu/pdf/Compare%20Sizing%20Methods.pdf (accessed April 27, 2018).

Horiba Scientific. *Choosing the Concentration Range for DLS Size Measurement.* http://www.horiba.com/scientific/products/particle-characterization/educa-tion/sz-100/particle-size-by-dynamic-light-scattering-resources/choosing-concentration-for-dls-size-measurement/.

Malvern Panalytical: *Materials Talks. Polydispersity-what does it mean for DLS and chromatography?* http://www.materials-talks.com/blog/2014/10/23/polydisper-sity-what-does-it-mean-for-dls-and-chromatography/.

E.E. Michaelides, Brownian movement and thermophoresis of nanoparticles in liquids. *Int. J. Heat Mass Transfer.*, 2015, **81**, 179.

C. Muhlfeld, B. Rothen-Rutishauser, D. Vanhecke, F. Blank, P. Gehr, and M. Ochs, Visualization and quantitative analysis of nanoparticles in the respiratory tract by transmission electron microscopy. *Part. Fibre. Toxicol.*, 2007, **4**, 11.

J. Stetefeld, S. A. McKenna, and T. R. Patel, Dynamic light scattering: a practical guide and applications in biomedical sciences. *Biophys. Rev.*, 2016, **8**, 409.

N. Seow, Y. N. Tan, L. Y. L. Yung, and X. Su, DNA-directed assembly of nanogold dimers: a unique dynamic light scattering sensing probe for transcription factor detection. *Sci. Rep.*, 2015, **5**, 18293.

C. T. Vogelson and A. R. Barron, Particle size control and dependence on solution pH of carboxylate-alumoxane nanoparticles, *J. Non-Cryst. Solids*, 2001, **290**, 216

J. Yang, L. B. Alemany, J. Driver, J. D. Hartgerink, and A. R. Barron, Fullerene-derivatized amino acids: synthesis, characterization, antioxidant properties, and solid phase peptide synthesis. *Chem. Eur. J.*, 2007, **3**, 2530.

J. Yang, K. Wang, J. Driver, J. Yang, and A. R. Barron, The use of fullerene substi-tuted phenylalanine derivatives as a passport through cell membranes. *Org. Biomol. Chem.*, 2007, **5**, 260.

Chapter 11: Raman Spectroscopy

Richa Sethi, Zheng Yan and Andrew R. Barron

Characterization of single-walled carbon nanotubes by Raman spectroscopy

Carbon nanotubes (CNTs) have proven to be a unique system for the application of Raman spectroscopy, and at the same time Raman spectroscopy has provided an exceedingly powerful tool useful in the study of the vibrational properties and electronic structures of CNTs. Raman spectroscopy has been successfully applied for studying CNTs at single nanotube level.

The large van der Waals interactions between the CNTs lead to an agglomeration of the tubes in the form of bundles or ropes. This problem can be solved by wrapping the tubes in a surfactant or functionalizing the SWCNTs by attaching appropriate chemical moieties to the sidewalls of the tube. Functionalization causes a local change in the hybridization from sp^2 to sp^3 of the side-wall carbon atoms, and Raman spectroscopy can be used to determine this change. In addition, information on length, diameter, electronic type (metallic or semiconducting), and whether nanotubes are separated or in bundle can be obtained by the use of Raman spectroscopy. Recent progress in understanding the Raman spectra of single walled carbon nanotubes (SWCNTs) have stimulated Raman studies of more complicated multi-wall carbon nanotubes (MWCNTs), but unfortunately quantitative determination of the latter is not possible at the present state of art.

Characterizing SWCNTs

Raman spectroscopy is a single resonance process, i.e., the signals are greatly enhanced if either the incoming laser energy (E_{laser}) or the scattered radiation matches an allowed electronic transition in the sample. For this process to occur, the phonon modes are assumed to occur at the center of the Brillouin zone (q = 0). Owing to their one-dimensional nature, the Π-electronic density of states of a perfect, infinite, SWCNTs form sharp singularities which are known as van Hove singularities (vHs), which are energetically symmetrical with respect to Fermi level (E_f) of the individual SWCNTs. The allowed optical transitions occur between matching vHs of the valence and conduction band of the SWCNTs, i.e., from first valence band vHs to the first conduction band vHs (E_{11}) or from the second vHs of the valence band to the second vHs of the conduction band (E_{22}). Since the quantum state of an electron (k) remains the same during the transition, it is referred to as k-selection rule.

The electronic properties, and therefore the individual transition energies in SWCNTs are given by their structure, i.e., by their chiral vector that determines the way SWNT is rolled up to form a cylinder. Figure 11.1 shows a SWCNT having vector R making an angle θ, known as the chiral angle, with the so-called zigzag or r_1 direction.

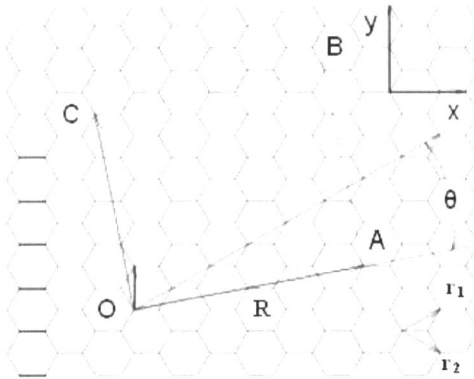

Figure 11.1: The unrolled honeycomb lattice of a nanotube. When the sites O and A, and the sites B and C are connected, a portion of a graphene sheet can be rolled seamlessly to form a SWCNT. The vectors OA and OB define the chiral vector R of the nanotube, respectively. The rectangle OABC defines the unit cell if the nanotube. The figure is constructed for $(n,m) = (4,2)$ nanotube. Adapted from M. S. Dresselhaus, G. Dresselhaus, R. Saito, and A. Jorio, Raman spectroscopy of carbon nanotubes. *Phys. Rep.*, 2004, 2, 47. Copyright: Elsevier (2004).

Raman spectroscopy of an ensemble of many SWCNTs having different chiral vectors is sensitive to the subset of tubes where the condition of allowed transition is fulfilled. A 'Kataura-Plot' gives the allowed electronic transition energies of individual SWCNTs as a function of diameter d, hence information on which tubes are resonant for a given excitation wavelength can be inferred. Since electronic transition energies vary roughly as $1/d$, the question whether a given laser energy probes predominantly semiconducting or metallic tubes depends on the mean diameter and diameter distribution in the SWCNTs ensemble. However, the transition energies that apply to an isolated

SWCNTs do not necessarily hold for an ensemble of interacting SWCNTs owing to the mutual van der Waals interactions.

Figure 11.2 shows a typical Raman spectrum from 100 to 3000 cm^{-1} taken of SWCNTs produced by catalytic decomposition of carbon monoxide (HiPco-process). The two dominant Raman features are the radial breathing mode (RBM) at low frequencies and tangential (G-band) multi-feature at higher frequencies. Other weak features, such as the disorder induced D-band and the G' band (an overtone mode) are also shown.

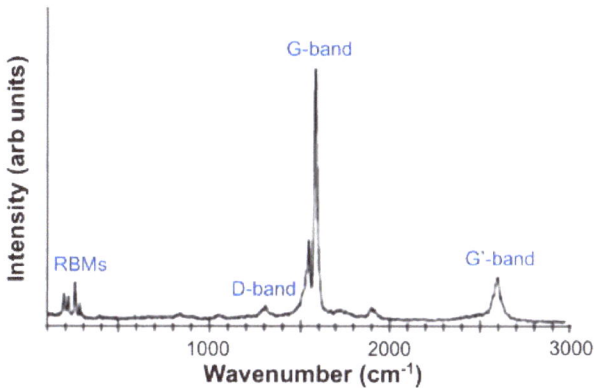

Figure 11.2: Raman spectrum of HiPco SWCNTs using a laser of wavelength of λ_{exc} = 633 nm. Adapted from R. Graupner, Raman spectroscopy of covalently functionalized single-wall carbon nanotubes. *J. Raman Spectrosc.*, 2007, 38, 673. Copyright: John Wiley & Sons (2007).

Modes in the Raman spectra of SWCNTs

Radial breathing modes (RBMs)

Out of all Raman modes observed in the spectra of SWCNTs, the radial breathing modes are unique to SWCNTs. They appear between 150 cm^{-1} < ω_{RBM} < 300 cm^{-1} from the elastically scattered laser line. It corresponds to the vibration of the C atoms in the radial direction, as if the tube is breathing (Figure 11.3). An important point about these modes is the fact that the energy (or wavenumber) of these vibrational modes depends on the diameter (d) of the SWCNTs, and not on the way the SWCNT is rolled up to form a cylinder, i.e., they do not depend on the θ of the tube.

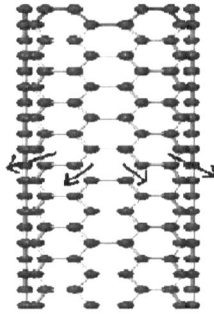

Figure 11.3: Schematic picture showing RBM vibration for a SWCNT.
Adapted from A. Jorio, M. A. Pimenta, A. G. S. Filho, R. Saito, G. Dresselhaus,
and M. S. Dresselhaus, Characterizing carbon nanotube samples with reso-
nance Raman scattering. *New J. Phys.*, 2003, 5, 139. Copyright: Institute of
Physics (2003).

These features are very useful for characterizing nanotube diameters through the relation $\omega_{RBM} = A/d + B$, where A and B are constants and their variations are often attributed to environmental effects, i.e., whether the SWCNTs are present as individual tubes wrapped in a surfactant, isolated on a substrate surface, or in the form of bundles. However, for typical SWCNT bundles in the diameter range, $d = 1.5 \pm 0.2$ nm, A = 234 cm^{-1} nm and B = 10 cm^{-1} (where B is an upshift coming from tube-tube interactions). For isolated SWNTs on an oxidized Si substrate, A= 248 cm^{-1} nm and B = 0. As can be seen from Figure 11.4, the relation $\omega_{RBM} = A/d + B$ holds true for the usual diameter range i.e., when d lies between 1 and 2 nm. However, when d is less than 1 nm, nanotube lattice distortions lead to chirality dependence of ω_{RBM} and for large diameters tubes when, d is more than 2 nm the intensity of RBM feature is weak and is hardly observable.

Hence, a single Raman measurement gives an idea of the tubes that are in resonance with the laser line but does not give a complete characterization of the diameter distribution of the sample. However, by taking Raman spectra using many laser lines, a good characterization of the diameter distributions in the sample can be obtained. Also, natural line widths observed for isolated SWCNTs are $\omega_{RBM} = 3$ cm^{-1} but as the tube diameter is increased, broadening is observed which is denoted by Γ_{RBM}. It has been observed that for $d > 2$ nm, $\Gamma_{RBM} > 20$ cm^{-1}. For SWNT bundles, the line width does not reflect Γ_{RMB}, it rather reflects an ensemble of tubes in resonance with the energy of laser.

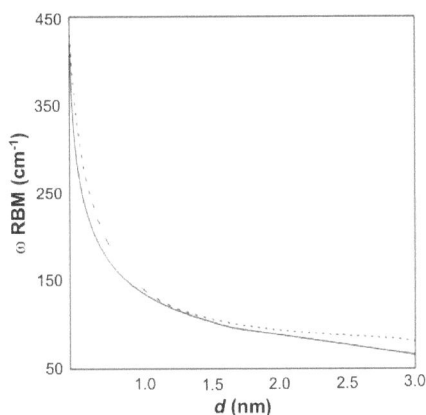

Figure 11.4: RBM frequencies $\omega_{RBM} = A/d + B$ versus nanotube diameter for (i) A = 234 cm^{-1} nm and B = 10 cm^{-1}, for SWCNT bundles (dashed curve); (ii) A = 248 cm^{-1} nm and B = 0, for isolated SWCNTs (solid curve). Adapted from A. Jorio, M. A. Pimenta, A. G. S. Filho, R. Saito, G. Dresselhaus, and M. S. Dresselhaus, Characterizing carbon nanotube samples with resonance Raman scattering. *New J. Phys.*, 2003, 5, 139. Copyright: Institute of Physics (2003).

Variation of RBM intensities upon functionalization

Functionalization of SWCNTs leads to variations of relative intensities of RBM compared to the starting material (unfunctionalized SWCNTs). Owing to the diameter dependence of the RBM frequency and the resonant nature of the Raman scattering process, chemical reactions that are sensitive to the diameter as well as the electronic structure, i.e., metallic or semiconducting of the SWCNTs can be sorted out. The difference in Raman spectra is usually inferred by thermal defunctionalization, where the functional groups are removed by annealing. The basis of using annealing for defunctionalizing SWCNTs is based on the fact that annealing restores the Raman intensities, in contrast to other treatments where a complete disintegration of the SWCNTs occurs. Figure 11.5 shows the Raman spectra of the pristine, functionalized and annealed SWCNTs. It can be observed that the absolute intensities of the radial breathing modes are drastically reduced after functionalization. This decrease can be attributed to vHs, which themselves are a consequence of translational symmetry of the SWCNTs. Since the translational symmetry of the SWCNTs is broken as a result of irregular distribution of the sp^3-sites due to the functionalization, these vHs are broadened and strongly reduced in intensity. As a result, the resonant Raman cross section of all modes is strongly reduced as well.

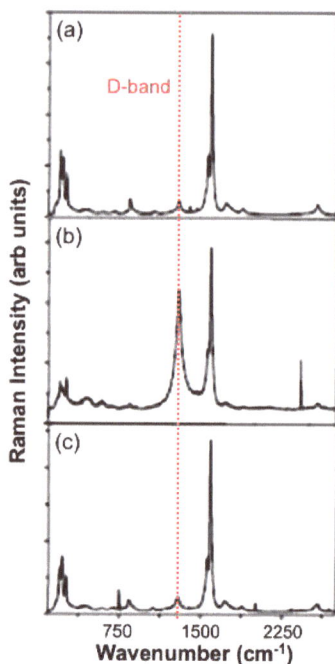

Figure 11.5: Raman spectra of SWCNT samples showing the effect of sidewall functionalization on the D-band: (a) pristine SWCNTs, (b) functionalized SWNTs, and (c) functionalized SWNTs after annealing at 750 °C in Ar. Adapted from R. Graupner, Raman spectroscopy of covalently functionalized single-wall carbon nanotubes. *J. Raman Spectrosc.*, **2007, 38, 673. Copyright: John Wiley & Sons (2007).**

For an ensemble of functionalized SWNTs, a decrease in high wavenumber RBM intensities has been observed which leads to an inference that destruction of small diameter SWNT takes place. Also, after prolonged treatment with nitric acid and subsequent annealing in oxygen or vacuum, diameter enlargement of SWCNTs is observed from the disappearance of RBMs from small diameter SWCNTs and the appearance of new RBMs characteristic of SWCNTs with larger diameters. In addition, laser irradiation seems to damage preferentially small diameter SWCNTs. In all cases, the decrease of RBM intensities is either attributed to the complete disintegration of SWCNTs or reduction in resonance enhancement of selectively functionalized SWCNTs. However, change in RBM intensities can also have other reasons. One reason is doping induced bleaching of electronic transitions in SWCNTs. When a dopant is added, a previously occupied electronic state can be filled or emptied, as a result of which E_f in the SWCNTs is shifted. If this shift is large enough and the conduction band vHs corresponding to the respective E_{ii} transition that is excited by the laser light gets occupied (n-type doping) or the valence band vHs is emptied (p-type doping), the resonant enhancement is lost as the electronic transitions are quenched.

Sample morphology has also seen to affect the RBMs. The same unfunctionalized sample in different aggregation states gives rise to different spectra. This is because the transition energy, E_{ii} depends on the aggregation state of the SWCNTs.

Tangential modes (G-band)

The tangential modes are the most intensive high-energy modes of SWNTs and form the so-called G-band, which is typically observed at around 1600 cm^{-1}. For this mode, the atomic displacements occur along the circumferential direction (Figure 11.6). Spectra in this frequency can be used for SWCNT characterization, independent of the RBM observation. This multi-peak feature can, for example, also be used for diameter characterization, although the information provided is less accurate than the RBM feature, and it gives information about the metallic character of the SWCNTs in resonance with laser line.

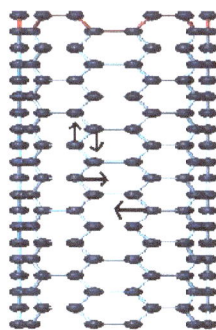

Figure 11.6: Schematic picture showing the atomic vibrations for the SWCNT G-band. Adapted from A. Jorio, M. A. Pimenta, A. G. S. Filho, R. Saito, G. Dresselhaus, and M. S. Dresselhaus, Characterizing carbon nanotube samples with resonance Raman scattering. *New J. Phys.*, 2003, 5, 139. Copyright: Institute of Physics (2003).

The tangential modes are useful in distinguishing semiconducting from metallic SWCNTs. The difference is evident in the G- feature (Figure 11.7 and Figure 11.8) which broadens and becomes asymmetric for metallic SWCNTs in comparison with the Lorentzian lineshape for semiconducting tubes, and this broadening is related to the presence of free electrons in nanotubes with metallic character. This broadened G-feature is usually fit using a Breit-Wigner-Fano (BWF) line that accounts for the coupling of a discrete phonon with a continuum related to conduction electrons. This BWF line is observed in many graphite-like materials with metallic character, such as n-doped graphite intercalation compounds (GIC), n-doped fullerenes, as well as metallic SWCNTs. The intensity of this G- mode depends on the size and number of metallic SWCNTs in a bundle (Figure 11.9).

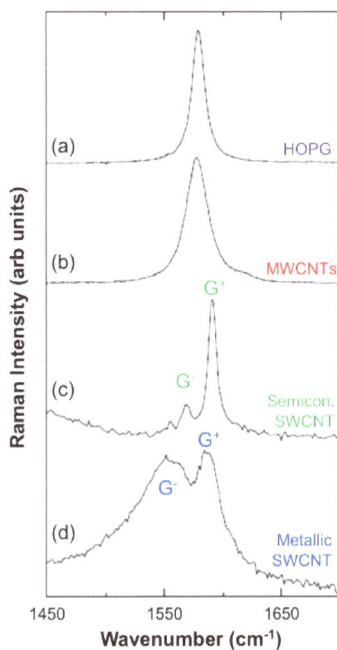

Figure 11.7: G-band for (a) highly ordered pyrolytic graphite (HOPG), (b) MWCNT bundles, (c) one isolated semiconducting SWCNT, and (d) one isolated metallic SWCNT. The multi-peak G-band feature is not clear for MWCNTs due to the large tube size. A. Jorio, M. A. Pimenta, A. G. S. Filho, R. Saito, G. Dresselhaus, and M. S. Dresselhaus, Characterizing carbon nanotube samples with resonance Raman scattering. *New J. Phys.*, 2003, 5, 139. Copyright Institute of Physics (2005).

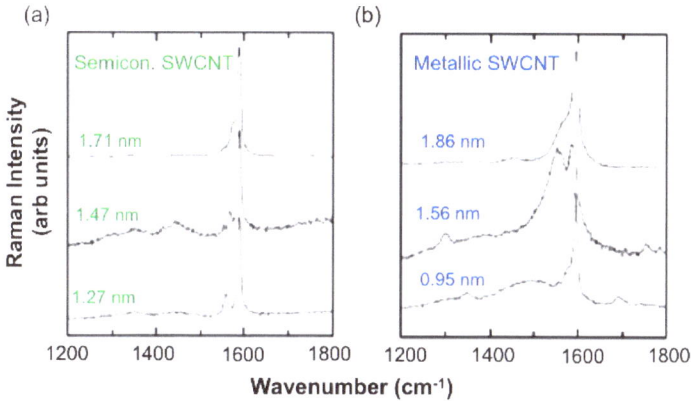

Figure 11.8: Raman signal from three isolated semiconducting and three isolated metallic SWCNTs showing the G-and D-band profiles. SWCNTs in good resonance (strong signal with low signal to noise ratio) show practically no D-band. Adapted from A. Jorio, M. A. Pimenta, A. G. S. Filho, R. Saito, G. Dresselhaus, and M. S. Dresselhaus, Characterizing carbon nanotube samples with resonance Raman scattering. *New J. Phys.*, 2003, 5, 139. Copyright Institute of Physics (2005).

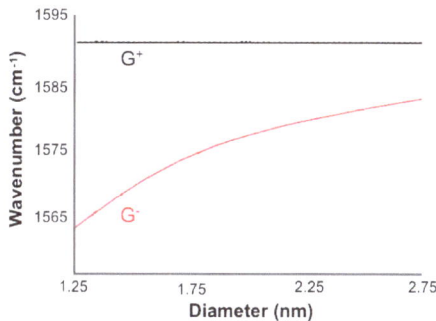

Figure 11.9: Dependence of G+ (black line) and G- (red line) frequencies as a function of diameter. Adapted from M. Paillet, T. Michel, J. C. Meyer, V. N. Popov, L. Henrad, S. Roth, and J. L. Sauvajol, Raman active phonons of identified semiconducting single-walled carbon nanotubes. *Phys. Rev. Lett.*, 2006, 96, 257401. Copyright: American Physical Society (2006).

Change of G-band line shape on functionalization

Chemical treatments are found to affect the line shape of the tangential line modes. Selective functionalization of SWNTs or a change in the ratio of metallic to semiconducting SWNTs due to selective etching is responsible for such a change. According to Figure 11.10, it can be seen that an increase or decrease of the BWF line shape is observed depending on the laser wavelength. At λ_{exc} = 633 nm, the preferentially functionalized small diameter SWNTs are semiconducting, therefore the G-band shows a decrease in the BWG asymmetry. However, the situation is reversed at 514 nm, where small metallic tubes are probed. BWF resonance intensity of small bundles increases with bundle thickness, so care should be taken that the effect ascribed directly to functionalization of the SWNTs is not caused by the exfoliation of the previously bundles SWNT.

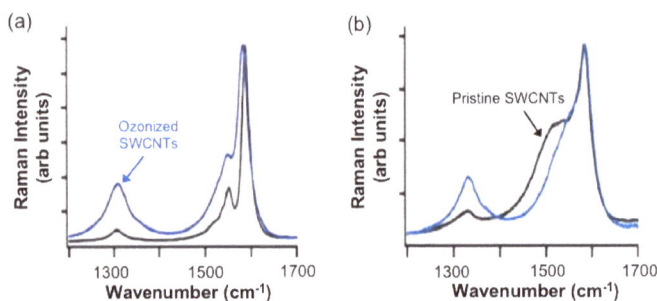

Figure 11.10: G-and D-band spectra of pristine (black) and ozonized (blue) SWNTs at (a) 633 nm and (b) 514 nm excitation. Adapted from R. Graupner, Raman spectroscopy of covalently functionalized single-wall carbon nanotubes. *J. Raman Spectrosc.*, 2007, 38, 673. Copyright: John Wiley & Sons (2007).

Disorder-induced D-band

This is one of the most discussed modes for the characterization of functionalized SWNTs and is observed at 1300-1400 cm^{-1}. Not only for functionalized SWNTs, D-band is also observed for unfunctionalized SWNTs. From a large number of Raman spectra from isolated SWNTs, about 50% exhibit observable D-band signals with weak intensity (Figure 11.8). A large D-peak compared with the G-peak usually means a bad resonance condition, which indicates the presence of amorphous carbon.

The appearance of D-peak can be interpreted due to the breakdown of the k-selection rule. It also depends on the laser energy and diameter of the SWNTs. This behavior is interpreted as a double resonance effect, where not only one of the direct, k-conserving electronic transitions, but also the emission of phonon is a resonant process. In contrast to single resonant Raman scattering, where only phonons around the center of the Brillouin zone (q = 0) are excited, the phonons that provoke the D-band exhibit a non-negligible q vector. This explains the double resonance theory for D-band in Raman spectroscopy. In few cases, the overtone of the D-band known as the G'-band (or D*-band) is observed at 2600-2800 cm^{-1}, and it does not require defect scattering as the two phonons with q and -q are excited. This mode is therefore observed independent of the defect concentration.

The presence of D-band cannot be correlated to the presence of various defects (such as hetero-atoms, vacancies, heptagon-pentagon pairs, kinks, or even the presence of impurities, etc.). Following are the two main characteristics of the D-band found in carbon nanotubes:

- Small linewidths: Γ_D values for SWNTs range from 40 cm^{-1} down to 7 cm^{-1}.
- Lower frequencies: D-band frequency is usually lower than the frequency of sp^2-based carbons, and this downshift of frequency shows $1/d$ dependence.

D-Band intensity as a measure of functionalization versus defect density

Since D-peak appears due to the presence defects, an increase in the intensity of the band is taken as a fingerprint for successful functionalization. But, whether D-band intensity is a measure of degree of functionalization or not is still sure. So, it is not correct to correlate D-peak intensity or D-peak area to the degree of functionalization. From Figure 11.11, it can be observed that for lower degree of functionalization, intensity of the D-band scales linearly with defect density. As the degree of functionalization is further increased, both D and G-band area decrease, which is explained by the loss of resonance enhancement due to functionalization. Also, normalization of the D-peak intensity to the G-band in order to correct for changes in resonance intensities also leads to a decrease for higher densities of functional groups.

Limitations of Raman spectroscopy

Though Raman spectroscopy has provided an exceedingly important tool for characterization of SWNTs, however, it suffers from few serious limitations. One of the main limitations of Raman spectroscopy is that it does not provide any information about the extent of functionalization in the SWNTs. The presence of D-band indicates disorder, i.e. side wall distribution, however it cannot differentiate between the number of substituents and their distribution. Following are the two main limitations of Raman spectroscopy.

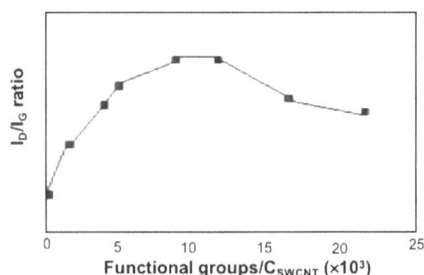

Figure 11.11: The intensity ratio I_D/I_G at λ_{exc} = 532 nm of functionalized SWCNTs with respect to degree of functionalization using diazonium reagents. Adapted from R. Graupner, Raman spectroscopy of covalently functionalized single-wall carbon nanotubes. *J. Raman Spectrosc.*, 2007, 38, 673. Copyright: John Wiley & Sons (2007).

Quantification of substituents

This can be illustrated by the following examples. Purified HiPco tubes may be fluorinated at 150 °C to give F-SWNTs with a C:F ratio of approximately 2.4:1. The Raman spectra (using 780 nm excitation) for F-SWNTs shows in addition to the tangential mode at ~1587 cm^{-1} an intense broad D (disorder) mode at ~ 1295 cm^{-1} consistent with the side wall functionalization. Irrespective of the arrangements of the fluorine substituents, thermolysis of F-SWNTs results in the loss of fluorine and the re-formation of unfunctionalized SWNTs alnog with their cleavage into shorter length tubes. As can be seen from Figure 11.12, the intensity of the D-band decreases as the thermolysis temperature increases. This is consistent with the loss of F-substituents. The G-band shows a concomitant sharpening and increase in intensity.

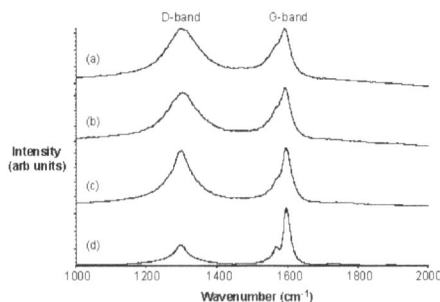

Figure 11.12: Raman spectra of F-SWNTs (a) as prepared at 150 °C and after heating to (b) 400, (c) 450 and (d) 550 °C.

As discussed above, the presence of a significant D mode has been the primary method for determining the presence of sidewall functionalization. It has been commonly accepted that the relative intensity of the D mode *versus* the tangential G mode is a quantitative measure of level of substitution. However, as discussed below, the G:D ratio is also dependent on the distribution of substituents. Using Raman spectroscopy in combination with XPS analysis of F-SWNTs that have been subjected to thermolysis at different temperatures, a measure of the accuracy of Raman as a quantitative tool for determining substituent concentration can be obtained. As can be seen from Figure 11.13, there is essentially no change in the G:D band ratio despite a doubling amount of functional groups. Thus, at low levels of functionalization the use of Raman spectroscopy to quantify the presence of fluorine substituents is a clearly suspect.

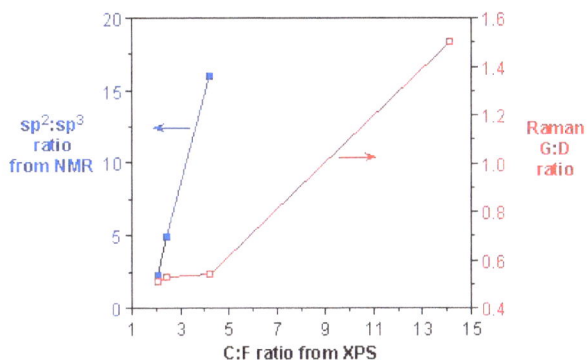

Figure 11.13: $C(sp^2):C-F(sp^3)$ ratio (blue) and Raman G-band:D-band ratio (red) as a function of C:F ratio from XPS.

On the basis of above data, it can be concluded that Raman spectroscopy does not provide an accurate quantification of small differences at low levels of functionalization, whereas when a comparison between samples with high levels of functionalization or large differences in degree of functionalization is requires Raman spectroscopy provides a good quantification.

Number versus distribution

Fluorinated nanotubes may be readily functionalized by reaction with the appropriate amine in the presence of base according to the scheme shown in Figure 11.14.

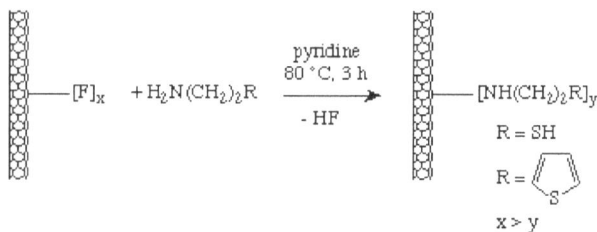

Figure 11.14: Synthesis of functionalized SWNTs.

When the Raman spectra of the functionalized SWNTs is taken (Figure 11.15), it is found out that the relative intensity of the disorder D-band at ~1290 cm^{-1} *versus* the tangential G-band (1500 - 1600 cm^{-1}) is much higher for thiophene-SWNT than thiol-SWNT. If the relative intensity of the D mode is the measure of the level of substitution, it can be concluded that there are a greater number of thiophene groups present per C than thiol groups. However, from the TGA weight loss data the SWNT-C:substituent ratios are calculated to be 19:1 and 17.5:1. Thus, contrary to the Raman data the TGA suggest that the number of substituents per C (in the SWNT) is actually similar for both substituents.

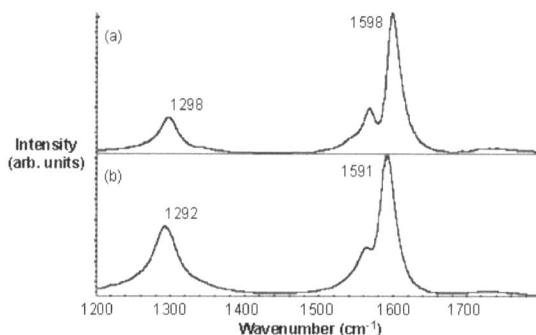

Figure 11.15: Raman spectrum of (a) thiol-SWNT and (b)thiophene-SWNT using 780 nm excitation showing the relative intensity of D-band at ~1300 cm^{-1} versus the G-band at ~1590 cm^{-1}.

This result would suggest that Raman spectroscopy is potentially unsuccessful in correctly providing the information about the number of substituents on the SWNTs. Subsequent imaging of the functionalized SWNTs by STM showed that the distribution of the functional groups was the difference between the thiol and thiphene functionalized SWNTs Figure 11.16. Thus, relative ratio of the D- and G-bands is a measure of concentration and distribution of functional groups on SWNTs.

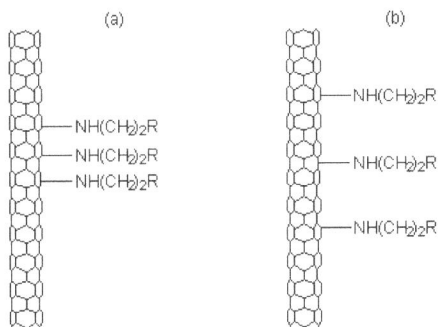

Figure 11.16: Schematic representation of the functional group distribution for (a) thiol-SWNT and (b) thiophene-SWNT.

Multi-walled carbon nanotubes (MWCNTs)

Most of the characteristic differences that distinguish the Raman spectra in SWCNTs from the spectra of graphite are not so evident for MWNTs. It is because the outer diameter for MWCNTs is very large and the ensemble of CNTs in them varies from small to very large. For example, the RBM Raman feature associated with a small diameter inner tube (less than 2 nm) can sometimes be observed when a good resonance condition is established, but since the RBM signal from large diameter tubes is usually too weak to be observable and the ensemble average of inner tube diameter broadens the signal, a good signal is not observed. However, when hydrogen gas in the arc discharge method is used, a thin innermost nanotube within a MWCNT of diameter 1 nm can be obtained which gives strong RBM peaks in the Raman spectra.

Whereas the G+ - G- splitting is large for small diameter SWNT, the corresponding splitting of the G-band in MWCNTs is both small in intensity and smeared out due to the effect of the diameter distribution. Therefore the G-band feature predominantly exists a weakly asymmetric characteristic line shape, and a peak appearing close to the graphite frequency of 1582 cm^{-1}.however for isolated MWCNTs prepared in the presence of hydrogen gas using the arc discharge method, it is possible to observe multiple G-band splitting effects even more clearly than for the SWNTs, and this is because environmental effects become relatively small for the innermost nanotube in a MWCNTs relative to the interactions occurring between SWCNTs and different environments. The Raman spectroscopy of MWCNTs has not been well investigated up to now. The new directions in this field are yet to be explored.

Characterization of graphene by Raman spectroscopy

Graphene is a quasi-two-dimensional material, which comprises layers of carbon atoms arranged in six-member rings (Figure 11.17). Since being discovered by Andre Geim (Figure 11.18) and co-workers at the University of Manchester, graphene has become one of the most exciting topics of research because of its distinctive band structure and physical properties, such as the observation of a quantum hall effect at room temperature, a tunable band gap, and a high carrier mobility.

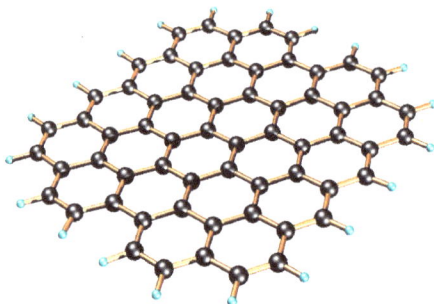

Figure 11.17: Idealized structure of a single graphene sheet. Copyright: Chris Ewels (http://www.www.ewels.info).

Figure 11.18: Russian-born Dutch-British physicist Sir Andre Konstantin Geim FRS (1958 -).

Graphene can be characterized by many techniques including atomic force microscopy (AFM), transmission electron microscopy (TEM) and Raman spectroscopy. AFM can be used to determine the number of the layers of the graphene, and TEM images can show the structure and morphology of the graphene sheets. In many ways, however, Raman spectroscopy is a much more important tool for the characterization of graphene. First of all, Raman spectroscopy is a simple tool and requires little sample preparation. What's more, Raman spectroscopy can not only be used to determine the number of layers, but also can identify if the structure of graphene is perfect, and if nitrogen, hydrogen or other functionalization is successful.

Raman spectrum of graphene

While Raman spectroscopy is a useful technique for characterizing sp^2 and sp^3 hybridized carbon atoms, including those in graphite, fullerenes, carbon nanotubes, and graphene. Single, double, and multi-layer graphenes have also been differentiated by their Raman fingerprints.

Figure 11.19 shows a typical Raman spectrum of N-doped single-layer graphene. The D-mode appears at approximately 1350 cm^{-1}, and the G-mode appears at approximately 1583 cm^{-1}. The other Raman modes are at 1620 cm^{-1} (D'- mode), 2680 cm^{-1} (2D-mode), and 2947 cm^{-1} (D+G-mode).

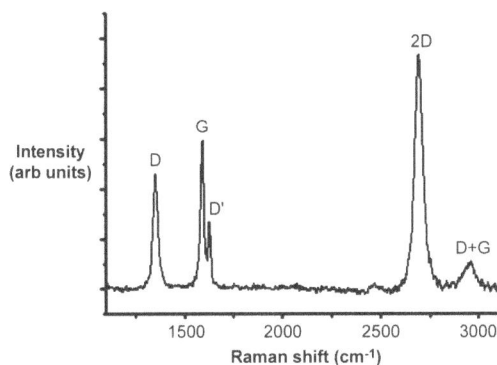

Figure 11.19: Raman spectrum of N-doped single-layer graphene with a 514.5 nm excitation laser wavelength.

The G-band

The G-mode is at about 1583 cm^{-1} and is due to E2g mode at the Γ-point. G-band arises from the stretching of the C-C bond in graphitic materials and is common to all sp^2 carbon systems. The G-band is highly sensitive to strain effects in sp^2 system, and thus can be used to probe modification on the flat surface of graphene.

Disorder-induced D- band and D'- band

The D-mode is caused by disordered structure of graphene. The presence of disorder in sp^2-hybridized carbon systems results in resonance Raman spectra, and thus makes Raman spectroscopy one of the most sensitive techniques to characterize disorder in sp^2 carbon materials. As is shown by a comparison of Figure 11.19 and Figure 11.20, there is no D peak in the Raman spectra of graphene with a perfect structure.

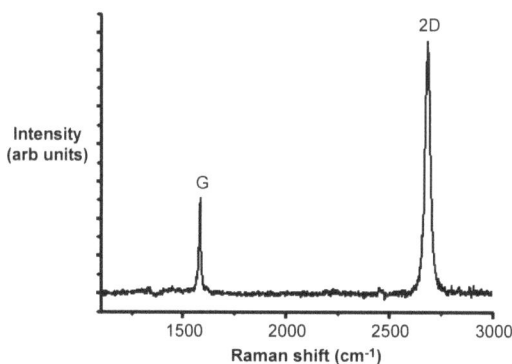

Figure 11.20: Raman spectrum with a 514.5 nm excitation laser wavelength of pristine single-layer graphene.

If there are some randomly distributed impurities or surface charges in the graphene, the G-peak can split into two peaks, G-peak (1583 cm^{-1}) and D'-peak (1620 cm^{-1}). The main reason is that the localized vibrational modes of the impurities can interact with the extended phonon modes of graphene resulting in the observed splitting.

The 2D-band

All kinds of sp^2 carbon materials exhibit a strong peak in the range 2500 - 2800 cm^{-1} in the Raman spectra. Combined with the G-band, this spectrum is a Raman signature of graphitic sp^2 materials and is called 2D-band. 2D-band is a second-order two-phonon process and exhibits a strong frequency dependence on the excitation laser energy.

What is more, the 2D band can be used to determine the number of layers of graphene. This is mainly because in the multi-layer graphene, the shape of 2D band is pretty much different from that in the single-layer graphene. As shown in Figure 11.21, the 2D band in the single-layer graphene is much more intense and sharper as compared to the 2D band in multi-layer graphene.

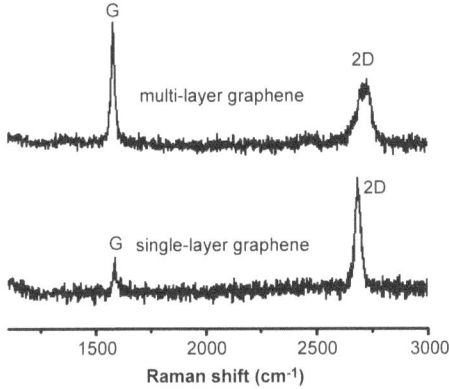

Figure 11.21: Raman spectrum with a 514.5 nm excitation laser wavelength of pristine single-layer and multi-layer graphene.

Bibliography

L. B. Alemany, L. Zhang, L. Zeng, C. L. Edwards, and A. R. Barron, *Chem. Mater.*, 2006, **19**, 735.

G. G. Chen, P. Joshi, S. Tadigadapa, and P. C. Eklund, Raman scattering from high-frequency phonons in supported *n*-graphene layer films. *Nano Lett.*, 2006, **6**, 2667.

Costa, B. Palen, M. Kruszynska, A. Bachmatiuk, and R.J. Kalenczuk, Characterization of carbon nanotubes by Raman spectroscopy. *Mat. Sci. Poland*, 2008, **26**, 433.

M. S. Dresselhaus, G. Dresselhaus, R. Saito, and A. Jorio, Raman spectroscopy of carbon nanotubes. *Phys. Rep.*, 2004, **2**, 47.

C. Ferrari, J. C. Meyer, V. Scardaci, C. Casiraghi, M. Lazzeri, F. Mauri, S. Piscanec, D. Jiang, K. S. Novoselov, and S. Roth, A. K. Geim, Raman spectrum of graphene and graphene layers. *Phys. Rev. Lett.*, 2006, **97**, 187401.

R. Graupner, Raman spectroscopy of covalently functionalized single-wall carbon nanotubes. *J. Raman Spectrosc.*, 2007, **38**, 673.

A. Jorio, M. A. Pimenta, A. G. S. Filho, R. Saito, G. Dresselhaus, and M. S. Dresselhaus, Characterizing carbon nanotube samples with resonance Raman scattering. *New J. Phys.*, 2003, **5**, 139.

M. Paillet, T. Michel, J. C. Meyer, V. N. Popov, L. Henrad, S. Roth, and J. L. Sauvajol, Raman active phonons of identified semiconducting single-walled carbon nanotubes. *Phys. Rev. Lett.*, 2006, **96**, 257401.

M. A. Pimenta, G. Dresselhaus, M. S. Dresselhaus, L. A.Cancado, A. Jorio, and R. Sato, Studying disorder in graphite-based systems by Raman spectroscopy. *Phys. Chem. Chem. Phys.*, 2007, **9**, 1276.

L. Zhang, J. Zhang, N. Schmandt, J. Cratty, V. N. Khabashesku, K. F. Kelly, and A. R. Barron, AFM and STM characterization of thiol and thiophene functionalized SWNTs: pitfalls in the use of chemical markers to determine the extent of sidewall functionalization in SWNTs. *Chem. Commun.*, 2005, 5429.

Chapter 12: NMR Spectroscopy

Amir Aliyan, Varun Shenoy Gangoli, Nadia Lara
and Andrew R. Barron

Introduction

Nuclear magnetic resonance (NMR) is the study of the nuclei of the response of an atom to an external magnetic field. Many nuclei have a net magnetic moment with $I \neq 0$, along with an angular momentum in one direction where I is the spin quantum number of the nucleus. In the presence of an external magnetic field, a nucleus would precess around the field. With all the nuclei precessing around the external magnetic field, a measurable signal is produced.

NMR can be used on any nuclei with an odd number of protons or neutrons or both, like the nuclei of hydrogen (^{1}H), carbon (^{13}C), phosphorous (^{31}P), etc. Hydrogen has a relatively large magnetic moment ($\mu = 14.1 \times 10^{-27}$ J/T) and hence it is used in NMR logging and NMR rock studies. The hydrogen nucleus composes of a single positively charged proton that can be seen as a loop of current generating a magnetic field. It is may be considered as a tiny bar magnet with the magnetic axis along the spin axis itself as shown Figure 12.1. In the absence of any external forces, a sample with hydrogen alone will have the individual magnetic moments randomly aligned as shown in Figure 12.2.

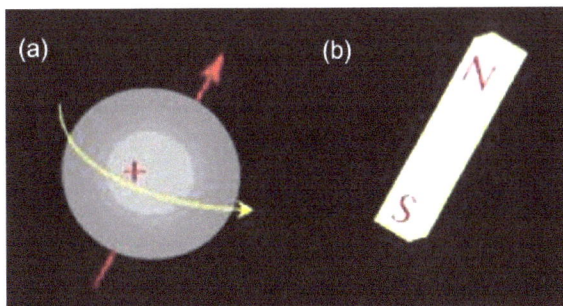

Figure 12.1: A simplistic representation of (a) a spinning nucleus as (b) a bar magnet. Copyright: Halliburton Energy Services, Duncan, OK (1999).

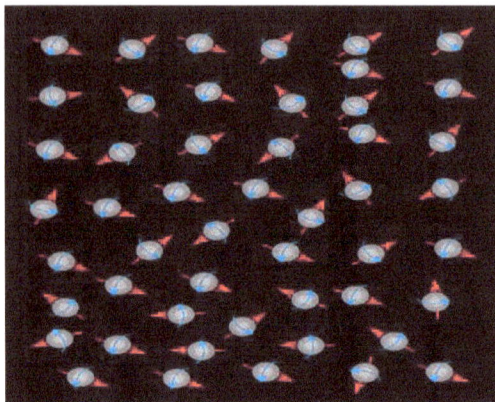

Figure 12.2: Representation of randomly aligned hydrogen nuclei. Copyright: Halliburton Energy Services, Duncan, OK (1999).

Measuring the specific surface area of nanoparticle suspensions using NMR

Surface area is a property of immense importance in the nano world, especially in the area of heterogeneous catalysis. A solid catalyst works with its active sites binding to the reactants, and hence for a given active site reactivity, the higher the number of active sites available, the faster the reaction will occur. In heterogeneous catalysis, if the catalyst is in the form of spherical nanoparticles, most of the active sites are believed to be present on the outer surface. Thus, it is very important to know the catalyst surface area in order to get a measure of the reaction time. One expresses this in terms of volume specific surface area, i.e., surface area/volume although in industry it is quite common to express it as surface area per unit mass of catalyst, e.g., m^2/g.

Advantages of NMR over BET

BET measurements follow the BET (Brunner-Emmet-Teller) adsorption isotherm of a gas on a solid surface. Adsorption experiments of a gas of known composition can help determine the specific surface area of the solid particle. This technique has been the main source of surface area analysis used industrially for a long time. However, BET techniques take a lot of time for the gas-adsorption step to be complete while one shall see in the course of this module that NMR can give you results in times averaging around 30 minutes depending on the sample. BET also requires careful sample preparation with the sample being in dry powder form, whereas NMR can accept samples in the liquid state as well.

How does NMR work?

Polarization

Polarization involves the alignment of the individual magnetic nuclei in the presence of a static external magnetic field B_o. This external field exerts a torque that forces the spinning nuclei to precess around it by a frequency given by the *Larmor frequency* given by,

$$f = \frac{\gamma B_o}{2\pi}$$

where γ is the gyromagnetic ratio which is a characteristic property of the nucleus. For hydrogen, $\gamma/2\pi = 42.58$ MHz/Tesla. This value is different for different elements.

Considering the case of a proton under the influence of an external magnetic field, it will be in one of two possible energy states depending on the orientation of the precession axis. If the axis is parallel to B_o, the proton is in the lower energy state (preferred state) and in the higher energy state if anti-parallel shown by Figure 12.3.

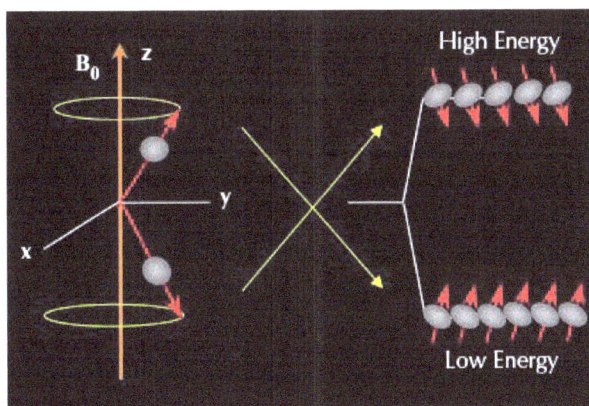

Figure 12.3: Alignment of precession axis. Copyright: Halliburton Energy Services, Duncan, OK (1999).

We define the net magnetization per unit volume of material, M_o, from Curie's law as,

$$M_o = \frac{N\gamma^2 h^2 I (I+1) B_o}{12\Pi^2 kT}$$

where N = number of protons, h = Planck's constant (6.626×10^{-34} Js), I = spin quantum number of the nucleus, k = Boltzmann's constant (1.381×10^{-23} m^2 Kg s^{-2} K^{-1}), and T = temperature (K).

T_1 relaxation

The protons are said to be polarized completely once they are all aligned with the static external field. Polarization grows with a time constant called the *longitudinal relaxation time* (T_1) as shown in,

$$M_z(t) = M_0(1 - e^{-t/T_1})$$

where t = time of exposure to B_o, $M_z(t)$ = magnitude of magnetization at time t, with B_o along z-axis and T_1 = time at which $M_z(t)$ reaches 90% of its final value, i.e., M_o.

T_1 is the time at which $M_z(t)$ reaches 63% of its final value, M_o. A typical T_1 relaxation experiment involves application of a 90° RF pulse that rotates the magnetization to the transverse direction. With time, the magnetization returns to its original value in the same fashion described by the above equation.

T_2 Relaxation

Once the polarization is complete, the magnetization direction is tipped from the longitudinal plane to a transverse plane by applying an oscillating field B_1 perpendicular to B_o. The frequency of B_1 must equal the Larmor frequency of the material from B_o. This oscillating field causes a possible change in energy state, and in-phase precession. The total phenomenon is called *nuclear magnetic resonance* as shown in Figure 12.4.

The oscillating field is generally pulsed in nature and so terms in books such a 180° pulse or 90° pulse indicates the angle through which the net magnetization gets tipped over. Application of a 90° pulse causes precession in the transverse phase. When the field B_1 is removed, the nuclei begin to de-phase and the net magnetization decreases. Here a receiver coil detects the decaying signal in a process called *free induction decay (FID)*. This exponential decay has an FID time constant (T_2) which is in the order of microseconds.

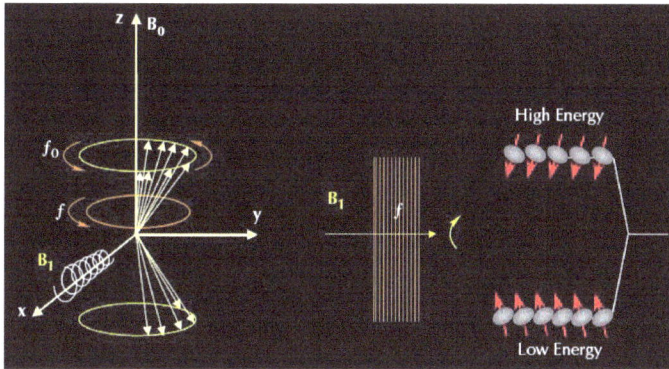

Figure 12.4: A schematic representation of the phenomenon of nuclear magnetic resonance. Copyright: Halliburton Energy Services, Duncan, OK (1999).

The time constant of the transverse relaxation is referred as T_2, and the amplitude of the decaying signal is given by,

$$M_x(t) = M_0(e^{-t/T_2})$$

with symbols as defined earlier.

CPMG sequence

The de-phasing caused by T_1 relaxation can be reversed by applying a 180° pulse after a time τ has passed after application of the initial 90° pulse. Thus, the phase of the transverse magnetization vector is now reversed by 180° so that "slower" vectors are now ahead of the "faster" vectors. These faster vectors eventually over-take the slower vectors and cause rephasing which is detected by a receiver coil as a *spin echo*. Thus, time τ also passes between the application of the 180° pulse and the maximum peak in the spin echo. The entire sequence is illustrated in Figure 12.5. A single echo decays very quickly and hence a series of 180° pulses are applied repeatedly in a sequence called the *Carr-Purcell-Meiboom-Gill (CPMG) sequence*.

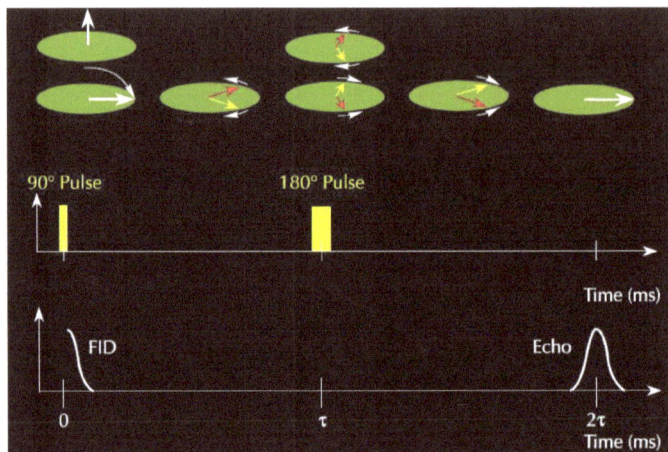

Figure 12.5: A schematic representation of the generation of a spin echo. Copyright: Halliburton Energy Services, Duncan, OK (1999).

NMR relaxation mechanism in solid suspensions

Calculations

From an atomic standpoint, T_1 relaxation occurs when a precessing proton transfers energy with its surroundings as the proton relaxes back from higher energy state to its lower energy state. With T_2 relaxation, apart from this energy transfer there is also dephasing and hence T_2 is less than T_1 in general. For solid suspensions, there are three independent relaxation mechanisms involved:

- Bulk fluid relaxation, which affects both T_1 and T_2 relaxation.
- Surface relaxation, which affects both T_1 and T_2 relaxation.
- Diffusion in the presence of the magnetic field gradients, which affects only T_2 relaxation.

These mechanisms act in parallel so that the net effects are given by,

$$\frac{1}{T_2} = \frac{1}{T_{2,bulk}} + \frac{1}{T_{2,surface}} + \frac{1}{T_{2,diffusion}}$$

and

$$\frac{1}{T_1} = \frac{1}{T_{1,bulk}} + \frac{1}{T_{1,surface}}$$

The relative importance of each of these terms depend on the specific scenario. For the case of most solid suspensions in liquid, the diffusion term can be ignored by having a relatively uniform external magnetic field that eliminates magnetic gradients. Theoretical analysis has shown that the surface relaxation terms can be written as,

$$\frac{1}{T_{1,surface}} = \rho_1 \left(\frac{S}{V}\right)_{particle}$$

and

$$\frac{1}{T_{2,surface}} = \rho_2 \left(\frac{S}{V}\right)_{particle}$$

where ρ = surface relaxivity and s/v = specific surface area. Thus, one can use T_1 or T_2 relaxation experiment to determine the specific surface area. We shall explain the case of the T_2 technique further as,

$$\frac{1}{T_2} = \frac{1}{T_{2,bulk}} + \rho_2 \left(\frac{S}{V}\right)_{particle}$$

One can determine T_2 by spin-echo measurements for a series of samples of known S/V values and prepare a calibration chart as shown in Figure 12.6, with the intercept as $1/T_{2,bulk}$ and the slope as ρ_2, , one can thus find the specific surface area of an unknown sample of the same material.

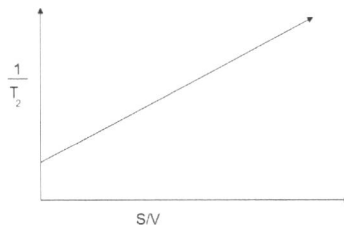

Figure 12.6: Example of a calibration plot of $1/T_2$ versus specific surface area (S/V) of a sample.

Sample preparation and experimental setup

The sample must be soluble in the solvent. For proton NMR, about 0.25 - 1.00 mg/mL are needed depending on the sensitivity of the instrument.

The solvent properties will have an impact of some or all of the spectrum. Solvent viscosity affects obtainable resolution, while other solvents like water or ethanol have exchangeable protons that will prevent the observation of such exchangeable protons present in the solute itself. Solvents must be chosen such that the temperature dependence of solute solubility is low in the operation temperature range. Solvents containing aromatic groups like benzene can cause shifts in the observed spectrum compared to non-aromatic solvents.

NMR tubes are available in a wide range of specifications depending on specific scenarios. The tube specifications need to be extremely narrow while operating with high strength magnetic fields. The tube needs to be kept extremely clean and free from dust and scratches to obtain good results, irrespective of the quality of the tube. Tubes can be cleaned without scratching by rinsing out the contents and soaking them in a degreasing solution, and by avoiding regular glassware cleaning brushes. After soaking for a while, rinse with distilled water and acetone and dry the tube by blowing filtered nitrogen gas through a pipette or by using a swab of cotton wool.

Filter the sample solution by using a Pasteur pipette stuffed with a piece of cotton wool at the neck. Any suspended material like dust can cause changes in the spectrum. When working with dilute aqueous solutions, sweat itself can have a major effect and so gloves are recommended at all times.

Sweat contains mainly water, minerals (sodium 0.9 g/L, potassium 0.2 g/L, calcium 0.015 g/L, magnesium 0.0013 g/L and other trace elements like iron, nickel, zinc, copper, lead and chromium), as well as lactate and urea. In presence of a dilute solution of the sample, the proton-containing substances in sweat (e.g., lactate and urea) can result in a large signal that can mask the signal of the sample.

The NMR probe is the most critical piece of equipment as it contains the apparatus that must detect the small NMR signals from the sample without adding a lot of noise. The size of the probe is given by the diameter of the NMR tube it can accommodate with common sizes 5, 10 and 15 mm. A larger size probe can be used in the case of less sensitive samples in order to get as much solute into the active zone as possible. When the sample is available in less quantity, use a smaller size tube to get an intrinsically higher sensitivity.

NMR analysis

A result sheet of T_2 relaxation has the plot of magnetization versus time, which will be linear in a semi-log plot as shown in Figure 12.7. Fitting it to the equation, we can find T_2 and thus one can prepare a calibration plot of $1/T_2$ versus S/V of known samples.

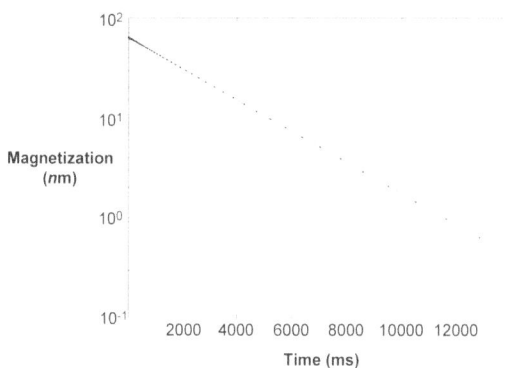

Figure 12.7: Example of T_2 relaxation with magnetization versus time on a semi-log plot.

Limitations of the T_2 technique

The following are a few of the limitations of the T_2 technique:

- One can't always guarantee no magnetic field gradients, in which case the T_1 relaxation technique is to be used. However, this takes much longer to perform than the T_2 relaxation.
- There is the requirement of the odd number of nucleons in the sample or solvent.
- The solid suspension should not have any para- or ferromagnetic substance (for instance, organics like hexane tend to have dissolved O_2 which is paramagnetic).
- The need to prepare a calibration chart of the material with known specific surface area.

Example of usage

A study of colloidal silica dispersed in water provides a useful example. Figure 12.8 shows a representation of an individual silica particle.

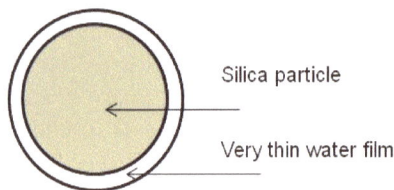

Figure 12.8: A representation of the silica particle with a thin water film surrounding it.

A series of dispersion in DI water at different concentrations was made and surface area calculated. The T_2 relaxation technique was performed on all of them with a typical T_2 plot shown in Figure 12.9 and T_2 was recorded at 2117 milliseconds for this sample.

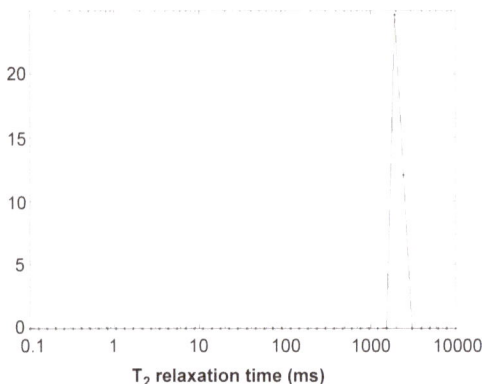

Figure 12.9: T_2 measurement for 2.3 wt% silica in DI water.

A calibration plot was prepared with $1/T_2 - 1/T_{2,bulk}$ as ordinate (the y-axis coordinate) and S/V as abscissa (the x-axis coordinate). This is called the *surface relaxivity plot* and is illustrated in Figure 12.10.

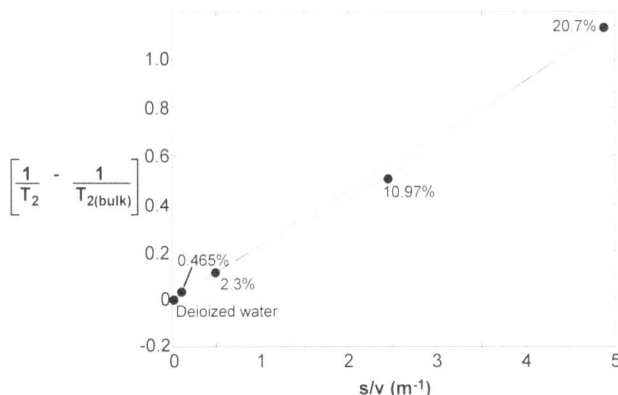

Figure 12.10: Calibration plot of $(1/T_2 - 1/T_{2,Bulk})$ versus specific surface area for silica in DI water.

Accordingly, for the colloidal dispersion of silica in DI water, the best fit resulted in,

$$\frac{1}{T_2} - \frac{1}{T_{2,bulk}} = 2.3x10^{-8}(\frac{S}{V}) - 0.0051$$

from which one can see that the value of surface relaxivity, 2.3×10^{-8}, is in close accordance with values reported in literature.

The T_2 technique has been used to find the pore-size distribution of water-wet rocks. Information of the pore size distribution helps petroleum engineers model the permeability of rocks from the same area and hence determine the extractable content of fluid within the rocks.

Usage of NMR for surface area determination has begun to take shape with a company, Xigo nanotools, having developed an instrument called the Acorn Area^{TM} to get surface area of a suspension of aluminum oxide. The results obtained from the instrument match closely with results reported by other techniques in literature. Thus, the T_2 NMR technique has been presented as a strong case to obtain specific surface areas of nanoparticle suspensions.

Using ^{13}C NMR to study carbon nanomaterials

There are several types of carbon nanomaterial. Members of this family are graphene, single-walled carbon nanotubes (SWNT), multi-walled carbon nanotubes (MWNT), and fullerenes such as C_{60}. Nano materials have been subject to various modification and functionalization, and it has been of interest to develop methods that could observe these changes. Herein we discuss selected applications of ^{13}C NMR in studying graphene and SWNTs. In addition, a discussion of how ^{13}C NMR could be used to analyze a thin film of amorphous carbon during a low-temperature annealing process will be presented.

^{13}C NMR versus ^{1}H NMR

Since carbon is found in any organic molecule NMR that can analyze carbon could be very helpful, unfortunately the major isotope, ^{12}C, is not NMR active. Fortunately, ^{13}C with a natural abundance of 1.1% is NMR active. This low natural abundance along with lower gyromagnetic ratio for ^{13}C causes sensitivity to decrease. Due to this lower sensitivity, obtaining a ^{13}C NMR spectrum with a specific signal-to-noise ratio requires averaging more spectra than the number of spectra that would be required to average in order to get the same signal to noise ratio for a ^{1}H NMR spectrum. Although it has a lower sensitivity, it is still highly used as it discloses valuable information.

Peaks in a ^{1}H NMR spectrum are split to $n + 1$ peak, where n is the number of hydrogen atoms on the adjacent carbon atom. The splitting pattern in ^{13}C NMR is different. First of all, C-C splitting is not observed, because the probability of having two adjacent ^{13}C is about 0.01%. Observed splitting patterns, which is due to the hydrogen atoms on the same carbon atom not on the adjacent carbon atom, is governed by the same $n + 1$ rule.

In ^{1}H NMR, the integral of the peaks are used for quantitative analysis, whereas this is problematic in ^{13}C NMR. The long relaxation process for carbon atoms takes longer comparing to that of hydrogen atoms, which also depends on the order of carbon (i.e., 1°, 2°, etc.). This causes the peak heights to not be related to the quantity of the corresponding carbon atoms.

Another difference between ^{13}C NMR and ^{1}H NMR is the chemical shift range. The range of the chemical shifts in a typical NMR represents the different between the minimum and maximum amount of electron density around that specific nucleus. Since hydrogen is surrounded by fewer electrons in comparison to carbon, the maximum change in the electron density for hydrogen is less than that for carbon. Thus, the range of chemical shift in ^{1}H NMR is narrower than that of ^{13}C NMR.

Solid state NMR

^{13}C NMR spectra could also be recorded for solid samples. The peaks for solid samples are very broad because the sample, being solid, cannot have all anisotropic, or orientation-dependent, interactions canceled due to *rapid random tumbling*. However, it is still possible to do high-resolution solid state NMR by spinning the sample at 54.74° with respect to the applied magnetic field, which is called the *magic angle*. In other words, the sample can be spun to artificially cancel the orientation-dependent interaction. In general, the spinning frequency has a considerable effect on the spectrum.

^{13}C NMR of carbon nanotubes

Single-walled carbon nanotubes contain sp^2 carbons. Derivatives of SWNTs contain sp^3 carbons in addition. There are several factors that affect the ^{13}C NMR spectrum of a SWNT sample, three of which will be reviewed in this module: [13]C percentage, diameter of the nanotube, and functionalization.

[13]C percentage

For sp^2 carbons, there is a slight dependence of ^{13}C NMR peaks on the percentage of ^{13}C in the sample. Samples with lower ^{13}C percentage are slighted shifted downfield (higher ppm). Data are shown in Table 12.1. Please note that these peaks are for the sp^2 carbons.

Sample	δ (ppm)
SWNTs (100%)	116 ± 1
SWNTs (1%)	118 ± 1

Table 12.1: Effects of ^{13}C percentage on the sp^2 peak. Data from S. Hayashi, F. Hoshi, T. Ishikura, M. Yumura, and S. Ohshima, ^{13}C NMR study of ^{13}C-enriched single-wall carbon nanotubes synthesized by catalytic decomposition of methane. *Carbon*, 2003, 41, 3047.

Diameter of the nanotubes

The peak position for SWNTs also depends on the diameter of the nanotubes. It has been reported that the chemical shift for sp^2 carbons decreases as the diameter of the nanotubes increases. Figure 12.11 shows this correlation. Since the peak position depends on the diameter of nanotubes, the peak broadening can be related to the diameter distribution. In other words, the narrower the peak is, the smaller the diameter distribution of SWNTs is. This correlation is shown in Figure 12.12.

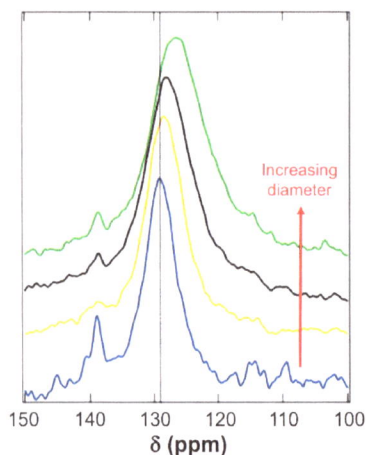

Figure 12.11: Correlation between the chemical shift of the sp^2 carbon and the diameter of the nanotubes as a function of increasing diameter. Adapted from C. Engtrakul, V. M. Irurzun, E. L. Gjersing, J. M. Holt, B. A. Larsen, D. E. Resasco, and J. L. Blackburn, Unraveling the ^{13}C NMR chemical shifts in single-walled carbon nanotubes: dependence on diameter and electronic structure. *J. Am. Chem. Soc.*, 2012, 134, 4850. Copyright: American Chemical Society (2012).

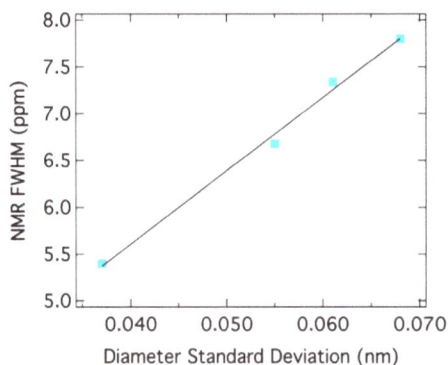

Figure 12.12: Correlation between ^{13}C NMR FWHM and the standard deviation of the diameter of nanotubes. Adapted from C. Engtrakul, V. M. Irurzun, E. L. Gjersing, J. M. Holt, B. A. Larsen, D. E. Resasco, and J. L. Blackburn, Unraveling the ^{13}C NMR chemical shifts in single-walled carbon nanotubes: dependence on diameter and electronic structure. *J. Am. Chem. Soc.*, 2012, 134, 4850. Copyright: American Chemical Society (2012).

Functionalization

Solid stated ^{13}C NMR can also be used to analyze functionalized nanotubes. As a result of functionalizing SWNTs with groups containing a carbonyl group, a slight shift toward higher fields (lower ppm) for the sp^2 carbons is observed. This shift is explained by the perturbation applied to the electronic structure of the whole nanotube as a result of the modifications on only a fraction of the nanotube. At the same time, a new peak emerges at around 172 ppm, which is assigned to the carboxyl group of the substituent. The peak intensities could also be used to quantify the level of functionalization. Figure 12.13 shows these changes, in which the substituents are $-(CH_2)_3COOH$, $-(CH_2)_2COOH$, and $-(CH_2)_2CONH(CH_2)_2NH_2$ for the spectra Figure 12.12b, Figure 12.13c, and Figure 12.12d, respectively. Note that the bond between the nanotube and the substituent is a C-C bond. Due to low sensitivity, the peak for the sp^3 carbons of the nanotube, which does not have a high quantity, is not detected. There is a small peak around 35 ppm in Figure 12.13, can be assigned to the aliphatic carbons of the substituent.

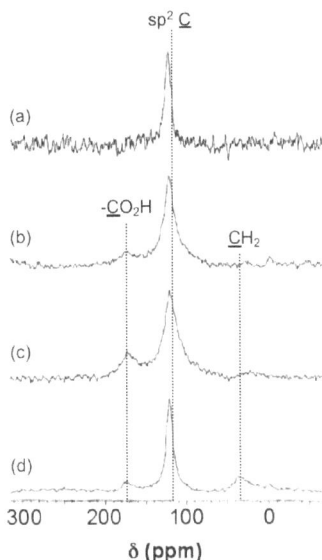

Figure 12.13: [13]**C NMR spectra for (a) pristine SWNT and SWNT functionalized with (b)** $-(CH_2)_3CO_2H$**, (c)** $-(CH_2)_2CO_2H$**, and (d)** $-(CH_2)_2CONH(CH_2)_2NH_2$**, showing the peak assignments. Adapted from H. Peng, L. B. Alemany, J. L. Margrave, and V. N. Khabashesku, Sidewall carboxylic acid functionalization of single-walled carbon nanotubes.** *J. Am. Chem. Soc.***, 2003, 125, 15174. Copyright: American Chemical Society (2003).**

For substituents containing aliphatic carbons, a new peak around 35 ppm emerges, as was shown in Figure 12.13, which is due to the aliphatic carbons. Since the quantity for the substituent carbons is low, the peak cannot be detected. Small substituents on the sidewall of SWNTs can be chemically modified to contain more carbons, so the signal due to those carbons could be detected. This idea, as a strategy for enhancing the signal from the substituents, can be used to analyze certain types of sidewall modifications. For example, when glycine (Gly, –NH$_2$CH$_2$CO$_2$H, Figure 12.14) was added to fluorinated SWNTs (F-SWNTs) to substitute the fluorine atoms, the ^{13}C NMR spectrum for the Gly-SWNTs was showing one peak for the sp^2 carbons. When the aliphatic substituent was changed to 6-aminohexanoic acid (Figure 12.15) with five aliphatic carbons, the peak was detectable, and using 11-aminoundecanoic acid (ten aliphatic carbons, Figure 12.16) the peak intensity was in the order of the size of the peak for sp^2 carbons. In order to use ^{13}C NMR to enhance the substituent peak (for modification quantification purposes as an example), Gly-SWNTs was treated with 1-dodecanol to modify Gly to an amino ester. This modification resulted in enhancing the aliphatic carbon peak at around 30 ppm. Similar to the results in Figure 12.13, a peak at around 170 emerged which was assigned to the carbonyl carbon. The sp^3 carbon of the SWNTs, which was attached to nitrogen, produced a small peak at around 80 ppm, which is detected in a cross-polarization magic angle spinning (CP-MAS) experiment.

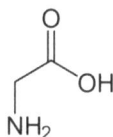

Figure 12.14: Structure of glycine (Gly, NH$_2$CH$_2$CO$_2$H).

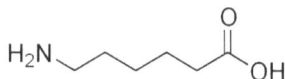

Figure 12.15: Structure of 6-aminohexanoic acid (NH$_2$(CH$_2$)$_5$CO$_2$H).

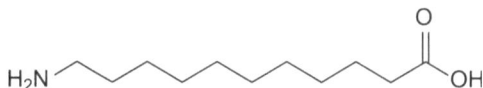

Figure 12.16: Structure of 11-aminoundecanoic acid (NH$_2$(CH$_2$)$_{10}$CO$_2$H).

F-SWNTs (fluorinated SWNTs) are reported to have a peak at around 90 ppm for the sp^3 carbon of nanotube that is attached to the fluorine. The results of this part are summarized in Table 12.2.

Group	δ (ppm)	Intensity
sp^2 carbons of SWNTs	120	Strong
$-NH_2(CH_2)_nCO_2H$ (aliphatic carbon, n=1,5, 10)	20-40	Depends on 'n'
$-NH_2(CH_2)_nCO_2H$ (carboxyl carbon, n=1,5, 10)	170	Weak
sp^3 carbon attached to nitrogen	80	Weak
sp^3 carbon attached to fluorine	90	Weak

Table 12.2: Chemical shift for different types of carbons in modified SWNTs. Note that the peak for the aliphatic carbons gets stronger if the amino acid is esterified. Data from H. Peng, L. B. Alemany, J. L. Margrave, and V. N. Khabashesku, Sidewall carboxylic acid functionalization of single-walled carbon nanotubes. *J. Am. Chem. Soc.*, 2003, 125, 15174; L. Zeng, L. Alemany, C. Edwards, and A. Barron, Demonstration of covalent sidewall functionalization of single wall carbon nanotubes by NMR spectroscopy: Side chain length dependence on the observation of the sidewall sp^3 carbons. *Nano. Res.*, 2008, 1, 72; L. B. Alemany, L. Zhang, L. Zeng, C. L. Edwards, and A. R. Barron, Solid-state NMR analysis of fluorinated single-walled carbon nanotubes: assessing the extent of fluorination. *Chem. Mater.*, 2007, 19, 735.

The peak intensities that are weak in Table 12.2 depend on the level of functionalization and for highly functionalized SWNTs, those peaks are not weak. The peak intensity for aliphatic carbons can be enhanced as the substituents get modified by attaching to other molecules with aliphatic carbons. Thus, the peak intensities can be used to quantify the level of functionalization.

^{13}C NMR of functionalized graphene

Graphene is a single layer of sp^2 carbons, which exhibits a benzene-like structure. Functionalization of graphene sheets results in converting some of the sp^2 carbons to sp^3. The peak for the sp^2 carbons of graphene shows a peak at around 140 ppm. It has been reported that fluorinated graphene produces a sp^3 peak at around 82 ppm. It has also been reported for graphite oxide (GO), which contains $-OH$ and epoxy substituents, to have peaks at around 60 and 70 ppm for the epoxy and the $-OH$ substituents, respectively. There are chances for similar peaks to appear for graphene oxide. Table 12.3 summarizes these results.

Type of carbon	δ (ppm)
sp^2	140
sp^3 attached to fluorine	80
sp^3 attached to –OH (for GO)	70
sp^3 attached to epoxide (for GO)	60

Table 12.3: Chemical shifts for functionalized graphene. Data from M. Dubois, K. Guérin, J. P. Pinheiro, Z. Fawal, F. Masin, and A. Hamwi, NMR and EPR studies of room temperature highly fluorinated graphite heat-treated under fluorine atmosphere. *Carbon*, 2004, 42, 1931; L. B. Casabianca, M. A. Shaibat, W. W. Cai, S. Park, R. Piner, R. S. Ruoff, and Y. Ishii, NMR-based structural modeling of graphite oxide using multidimensional ^{13}C solid-state NMR and ab initio chemical shift calculations. *J. Am. Chem. Soc.*, 2010, 132, 5672.

Analyzing annealing process using ^{13}C NMR

^{13}C NMR spectroscopy has been used to study the effects of low-temperature annealing (at 650 °C) on thin films of amorphous carbon. The thin films were synthesized from a ^{13}C enriched carbon source (99%). There were two peaks in the ^{13}C NMR spectrum at about 69 and 142 ppm which were assigned to sp^3 and sp^2 carbons, respectively. The intensity of each peak was used to find the percentage of each type of hybridization in the whole sample, and the broadening of the peaks was used to estimate the distribution of different types of carbons in the sample. It was found that while the composition of the sample didn't change during the annealing process (peak intensities didn't change, see Figure 12.17a), the full width at half maximum (FWHM) did change (Figure 12.17b). The latter suggested that the structure became more ordered, i.e., the distribution of sp^2 and sp^3 carbons within the sample became more homogeneous. Thus, it was concluded that the sample turned into a more homogenous one in terms of the distribution of carbons with different hybridization, while the fraction of sp^2 and sp^3 carbons remained unchanged.

Aside from the reported results from the paper, it can be concluded that ^{13}C NMR is a good technique to study annealing, and possibly other similar processes, in real time, if the kinetics of the process is slow enough. For these purposes, the peak intensity and FWHM can be used to find or estimate the fraction and distribution of each type of carbon respectively.

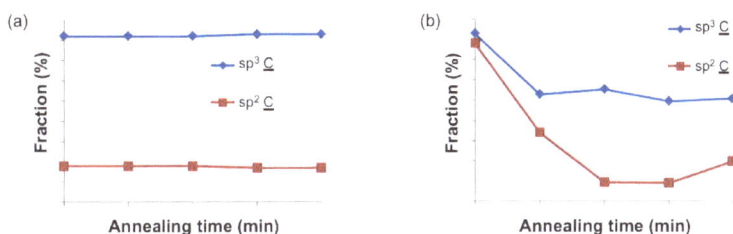

Figure 12.17: (a) Effect of the annealing process on the FWHM, which represents the change in the distribution of sp² and sp³ carbons. (b) Fractions of sp² and sp³ carbon during the annealing process. Data from T. M. Alam, T. A. Friedmann, P. A. Schultz, and D. Sebastiani, Low temperature annealing in tetrahedral amorphous carbon thin films observed by ^{13}C NMR spectroscopy. *Phys. Rev. B.*, 2003, 67, 245309. Copyright: American Physical Society (2003).

Measuring epoxide content of carbon nanomaterials.

One can measure the amount of epoxide on nanomaterials such as carbon nanotubes and fullerenes by monitoring a reaction involving phosphorus compounds in a similar manner to that described above. This technique uses the catalytic reaction of methyltrioxorhenium,

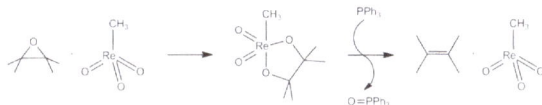

An epoxide reacts with methyltrioxorhenium ($MeReO_3$) to form a five membered ring. In the presence of triphenylphosphine (PPh_3), the catalyst is regenerated, forming an alkene and triphenylphosphine oxide ($OPPh_3$). The same reaction can be applied to carbon nanostructures and used to quantify the amount of epoxide on the nanomaterial,

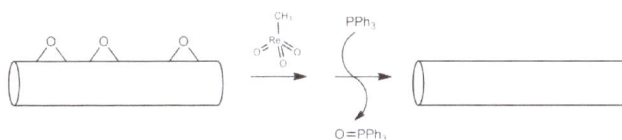

Figure 12.18 illustrates the quantification of epoxide on a carbon nanotube. Because the amount of initial PPh_3 used in the reaction is known, the relative amounts of PPh_3 and $OPPh_3$ can be used to stoichiometrically determine the

amount of epoxide on the nanotube. ^{31}P NMR spectroscopy is used to determine the relative amounts of PPh₃ and OPPh₃ (Figure 12.18).

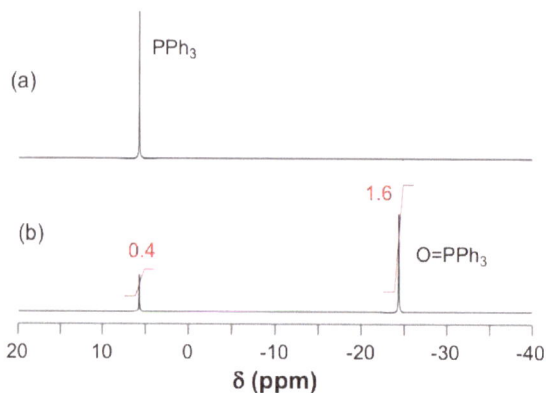

Figure 12.18: **[31]P spectrum of a mixture of PPh₃ and SWCNTs (a) before addition of MeReO₃ and (b) at oxygen transfer reaction is complete.**

The integration of the two ^{31}P signals is used to quantify the amount of epoxide on the nanotube according to,

$$\text{Moles of epoxide} = \frac{\text{area of OPPh}_3 \text{ peak}}{\text{area of PPh}_3 \text{ peak}} \times \text{moles PPh}_3$$

Thus, from a known quantity of PPh₃, one can find the amount of OPPh₃ formed and relate it stoichiometrically to the amount of epoxide on the nanotube. Not only does this experiment allow for such quantification, it is also unaffected by the presence of the many different species present in the experiment. This is because the compounds of interest, PPh₃ and OPPh₃, are the only ones that are characterized by ^{31}P NMR spectroscopy.

Bibliography

T. M. Alam, T. A. Friedmann, P. A. Schultz, and D. Sebastiani, Low temperature annealing in tetrahedral amorphous carbon thin films observed by ^{13}C NMR spectroscopy. *Phys. Rev. B.*, 2003, **67**, 245309

L. B. Alemany, L. Zhang, L. Zeng, C. L. Edwards, and A. R. Barron, Solid-state NMR analysis of fluorinated single-walled carbon nanotubes: assessing the extent of fluorination. *Chem. Mater.*, 2007, **19**, 735.

L. B. Casabianca, M. A. Shaibat, W. W. Cai, S. Park, R. Piner, R. S. Ruoff, and Y. Ishii, NMR-based structural modeling of graphite oxide using multidimensional ^{13}C solid-state NMR and ab initio chemical shift calculations. *J. Am. Chem. Soc.*, 2010, **132**, 5672.

J. Chattopadhyay, A. Mukherjee, C. E. Hamilton, J. H. Kang, S. Chakraborty, W. Guo, K. F. Kelly, A. R. Barron and W. E. Billups, Graphite epoxide. *J. Am. Chem. Soc.*, 2008, **130**, 5414.

G. R Coates, L. Xiao, and M.G. Prammer, *NMR Logging: Principles & Applications*, Halliburton Energy Services, Houston (2001).

B. Cowan, *Nuclear magnetic resonance and relaxation*, Cambridge University Press, Cambridge UK (2001).

A. E. Derome, *Modern NMR Techniques for Chemistry Research*, Vol 6, Pergamon Press, Oxford (1988).

M. Dubois, K. Guérin, J. P. Pinheiro, Z. Fawal, F. Masin, and A. Hamwi, NMR and EPR studies of room temperature highly fluorinated graphite heat-treated under fluorine atmosphere. *Carbon*, 2004, **42**, 1931;

C. Engtrakul, V. M. Irurzun, E. L. Gjersing, J. M. Holt, B. A. Larsen, D. E. Resasco, and J. L. Blackburn, Unraveling the ^{13}C NMR chemical shifts in single-walled carbon nanotubes: dependence on diameter and electronic structure. *J. Am. Chem. Soc.*, 2012, **134**, 4850.

S. Hayashi, F. Hoshi, T. Ishikura, M. Yumura, and S. Ohshima, ^{13}C NMR study of ^{13}C-enriched single-wall carbon nanotubes synthesized by catalytic decomposition of methane. *Carbon*, 2003, **41**, 3047.

W. E. Kenyon, Petrophysical principles of applications of NMR logging. *The Log Analyst*, 1997, **38**, SPWLA-1997-v38n2a4.

D. Ogrin, J. Chattopadhyay, A. K. Sadana, E. Billups, and A. R. Barron, Epoxidation and deepoxidation of single-walled carbon nanotubes: quantification of epoxide defects. *J. Am. Chem. Soc.*, 2006, **128**, 11322.

H. Peng, L. B. Alemany, J. L. Margrave, and V. N. Khabashesku, Sidewall carboxylic acid functionalization of single-walled carbon nanotubes. *J. Am. Chem. Soc.*, 2003, **125**, 15174.

L. Zeng, L. Alemany, C. Edwards, and A. Barron, Demonstration of covalent sidewall functionalization of single wall carbon nanotubes by NMR spectroscopy: Side chain length dependence on the observation of the sidewall sp^3 carbons. *Nano. Res.*, 2008, **1**, 72.

Chapter 13: Differential Mobility Analysis

Katherinne Requejo and Andrew R. Barron

Introduction

Electrospray-differential mobility analysis (ES-DMA) is an analytical technique that uses first an electrospray to aerosolize particles and then DMA to characterize their electrical mobility at ambient conditions. This versatile tool can be used to quantitative characterize biomolecules and nanoparticles from 0.7 to 800 nm. In the 1980s, it was discovered that ES could be used for producing aerosols of biomacromolecules. In the case of the DMA, its predecessor was developed by Hewitt in 1957 to analyze charging of small particles. The modified DMA, which is a type of ion mobility analyzer, was developed by Knutson and Whitby (Figure 13.1) in 1975 and later it was commercialized. Among the several designs, the cylindrical DMA has become the standard design and has been used for the obtention of monodisperse aerosols, as well as for the classification of polydisperse aerosols.

Figure 13.1: American engineer K. T. Whitby (1925 - 1983).

The first integration of ES with DMA occurred in 1996 when this technique was used to determine the size of different globular proteins. DMA was refined over the past decade to be used in a wide range of applications for the characterization of polymers, viruses, bacteriophages and nanoparticle-biomolecule conjugates. Although numerous publications have reported the use of ES-DMA in medicinal and pharmaceutical applications, this present module describes the general principles of the technique and its application in the analysis of gold nanoparticles.

How does ES-DMA function?

ES-DMA consists of an electrospray source (ES) that aerosolizes bionanoparticles and a class of ion mobility analyzer (DMA) that measures their electrical mobility by balancing electrical and drag forces on the particles. DMA continuously separates particles based on their charge to size ratio. A schematic of the experimental setup for ES-DMA is shown in Figure 13.2 for the analysis of gold nanoparticles.

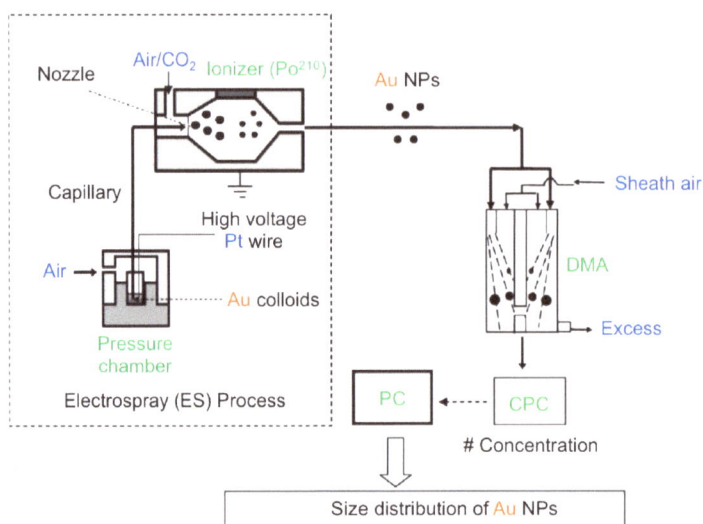

Figure 13.2: Schematic of experimental setup for ES-DMA. Adapted from D. Tsai, R. A. Zangmeister, L. F. Pease III, M. J. Tarlov, and M. R. Zachariah, Gas-phase ion-mobility characterization of SAM-functionalized Au nanoparticles. *Langmuir*, **2008, 24, 8483. Copyright: American Chemical Society (2015).**

The process of analyzing particles with ES-DMA involves four steps.

First, the analyte dissolved in a volatile buffer such as ammonium acetate $[NH_4][O_2CCH_3]$ is placed inside a pressure chamber. Then, the solution is delivered to the nozzle through a fused silica capillary to generate multiply charged droplets. ES nebulizers produce droplets of 100 - 400 nm in diameter but they are highly charged.

In the next step, the droplets are mixed with air and carbon dioxide (CO_2) and are passed through the charge reducer or neutralizer where the solvent continues to evaporate and charge distribution decreases. The charge reducer is an ionizing α radiation source such as Po^{210} that ionizes the carrier gas and reduces the net charges on the particles to a Fuchs'-Boltzmann distribution. As a result, the majority of the droplets contain single net charge particles that pass directly to the DMA. DMA separates positively or negatively charged particles by applying a negative or positive potential. Figure 13.3 shows a single channel design of cylindrical DMA that is composed of two concentric electrodes between which a voltage is applied. The inner electrode is maintained at a controlled voltage from 1V to 10 kV, whereas the outer electrode is electrically grounded.

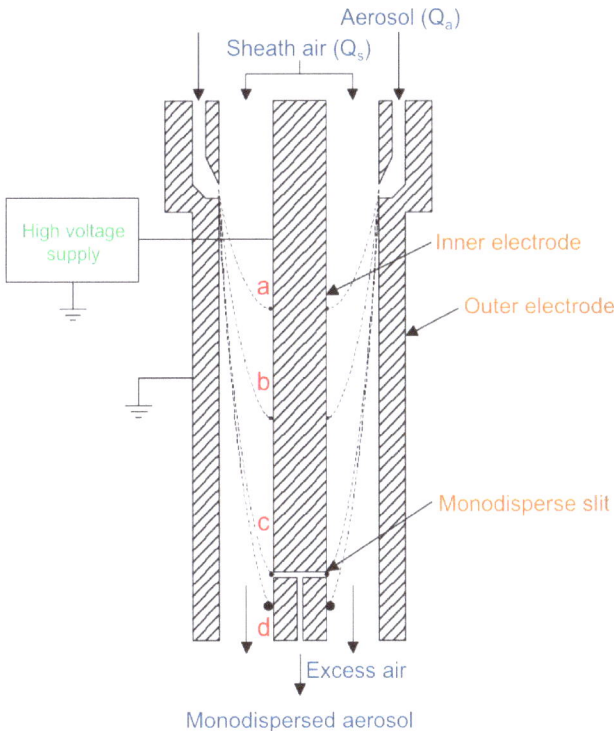

Figure 13.3: Basic principle of a general DMA. Adapted from P. Intra and N. Tippayawong, An overview of differential mobility analyzers for size classification of nanometer-sized aerosol particles. *Songklanakarin J. Sci. Technol.,* **2008, 30, 243. Copyright: Prince of Songkla University (2008).**

In the third step, the aerosol flow (Q_a) enters through a slit that is adjacent to one electrode and the sheath air (air or N_2) flow (Q_s) is introduced to separate the aerosol flow from the other electrode. After a voltage is applied between the inner and outer electrodes, an electric field is formed and the charged particles with specific electrical mobility are attracted to a charged collector rod. The positions of the charged particles along the length of the collector depend on their electrical mobility (Z_p), the fluid flow rate and the DMA geometry. In the case of particles with a high electrical mobility, they are collected in the upper part of the rod (particles a and b in Figure 13.3) while particles with a low electrical mobility are collected in the lower part of the rod (particle d, Figure 13.3).

The electrical mobility of the particles (Z_p) is a function of the fluid flow rate (Q_s and Q_a), the applied voltage (V) and the dimensions of the DMA as described by,

$$Z_p = \frac{(Q_s + Q_a)\ln(R_2/R_1)}{2\pi LV}$$

where R_1 and R_2 correspond to the radii of the outer and inner electrodes and L is the effective electrode length. Since DMA differentiates small changes in mobility, particles with a narrow range of electrical mobility (monodisperse aerosol) exit through a small slit at the bottom of the collector rod (particle c on inner electrode, Figure 13.3). The excess air flow contains remaining particles.

With the value of the electrical mobility, the particle diameter (d_p) can be determined by using Stokes' law as described by,

$$d_p = \frac{neC_c}{3\pi\mu Z_p}$$

where n is the number of charge units, e is the elementary unit of charge ($1.61 \times 10^{-19}C$), C_c is the Cunningham slip correction factor and μ is the gas viscosity. C_c considers the non-continuum flow effect when d_p is similar to or smaller than the mean free path (λ) of the carrier gas,

$$C_C = 1 + \frac{2\lambda}{d_p}[1.257 + 0.40\exp(\frac{-1.10d_p}{2\lambda})]$$

In the last step, the size-selected particles are detected with a condensation particle counter (CPC) or an aerosol electrometer (AE) that determines the particle number concentration. CPC has lower detection and quantitation limits and is the most sensitive detector available. AE is used when the particles concentrations are high or when particles are so small that cannot be detected by CPC. Figure 13.4 shows the operation of the CPC in which the aerosol is mixed with butanol (C_4H_9OH) or water vapor (working fluid) that condensates on the particles to produce supersaturation. Hence, large size particles (around 10 µm) are obtained, detected optically and counted. Since each droplet is approximately of the same size, the count is not biased. The particle size distribution is obtained by changing the applied voltage. Generally, the performance of the CPC is evaluated in terms of the minimum size that is counted with 50% efficiency.

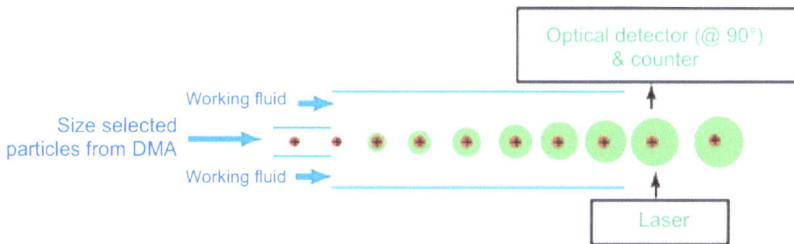

Figure 13.4: Working principle of the condensation particle counter (CPC). Adapted from S. Guha, M. Li, M. J. Tarlov, and M. R. Zachariah, Electrospray-differential mobility analysis of bionanoparticles, *Trends Biotechnol.*, 2012, 30, 291. Copyright: Elsevier (2015).

What type of information is obtained by ES-DMA?

ES-DMA provides information of the mobility diameter of particles and their concentration in number of particles per unit volume of analyzed gas so that the particle size distribution is obtained as shown in Figure 13.5. Another form of data representation is the differential distribution plot of $\Delta N/\Delta \log d_p$ versus d_P. This presentation has a logarithmic size axis that is usually more convenient because particles are often distributed over a wide range of sizes.

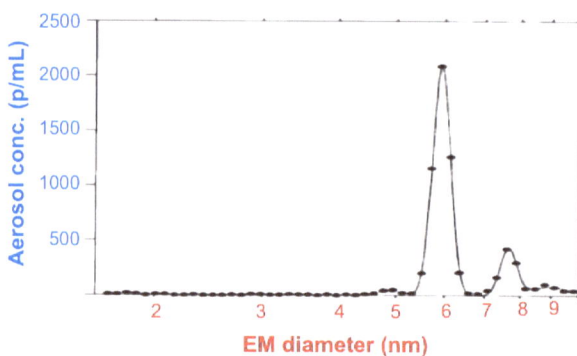

Figure 13.5: Size distribution of human serum albumin. Adapted from S. T. Kaufman, J. W. Skogen, F. D. Dorman, and F. Zarrin, Macromolecule analysis based on electrophoretic mobility in air: globular proteins. *Anal. Chem.*, 1996, 68, 1895. Copyright: American Chemical Society (2015).

How data from ES-DMA is processed?

To obtain the actual particle size distribution (Figure 13.5), the raw data acquired with the ES-DMA is corrected for charge correction, transfer function of the DMA and collection efficiency for CPC. Figure 13.6 illustrates the charge correction in which a charge reducer or neutralizer is necessary to reduce the problem of multiple charging and simplify the size distribution. The charge reduction depends on the particle size and multiple charging can be produced as the particle size increases. For instance, for 10 nm particles, the percentages of single charged particles are lower than those of neutral particles. After a negative voltage is applied, only the positive charged particles are collected. Conversely, for 100 nm particles, the percentages of single charged particles increase, and multiple charges are present. Hence, after a negative bias is applied, +1 and +2 particles are collected. The presence of more charges in particles indicates high electrical mobility and particles are observed at smaller diameter.

The transfer function for DMA modifies the input particle size distribution and affects the resolution as shown in Figure 13.7. This transfer function depends on the operation conditions such as flow rates and geometry of the DMA. Furthermore, the transfer function can be broadened by Brownian diffusion and this effect produces the actual size distribution. The theoretical resolution is measured by the ratio of the sheath to the aerosol flow in under balance flow conditions (sheath flow equals excess flow and aerosol flow in equals monodisperse aerosol flow out).

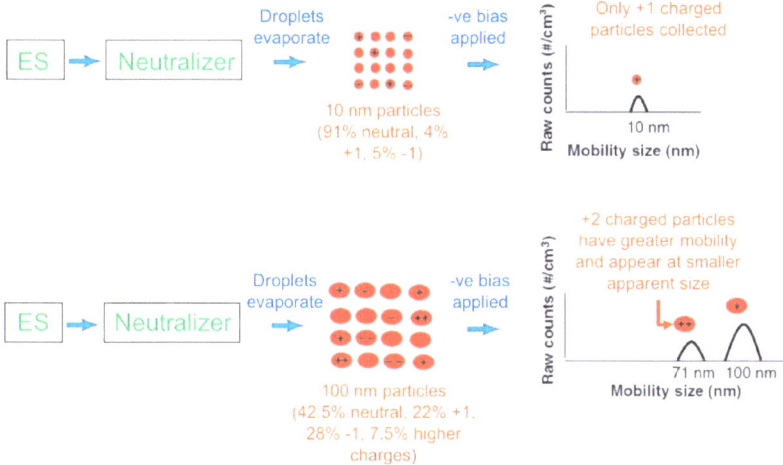

Figure 13.6: Data processing for the charge correction in the aerosol phase. Adapted from S. Guha, M. Li, M. J. Tarlov, and M. R. Zachariah, Electrospray-differential mobility analysis of bionanoparticles, *Trends Biotechnol.*, 2012, 30, 291. Copyright: Elsevier (2015).

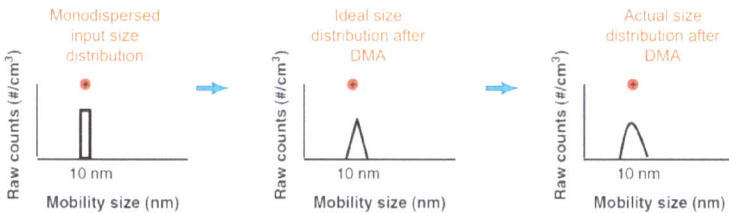

Figure 13.7: Data processing for transfer function for DMA. Adapted from S. Guha, M. Li, M. J. Tarlov, and M. R. Zachariah, Electrospray-differential mobility analysis of bionanoparticles, *Trends Biotechnol.*, 2012, 30, 291. Copyright: Elsevier (2015).

The CPC has a size limit of detection of 2.5 nm because small particles are difficult to activate at the supersaturation of the working fluid. Therefore, CPC collection efficiency is required that consists on the calibration of the CPC against an electrometer.

Applications of ES-DMA

- Determination of molecular weight of polymers and proteins in the range of 3.5 kDa to 2 MDa by correlating molecular weight and mobility diameter.
- Determination of absolute number concentration of nanoparticles in solution by obtaining the ES droplet size distributions and using statistical analysis to find the original monomer concentration. Dimers or trimers can be formed in the electrospray process due to droplet induced aggregation and are observed in the spectrum.
- Kinetics of aggregation of nanoparticles in solution by analysis of multimodal mobility distributions from which distinct types of aggregation states can be identified.
- Quantification of ligand adsorption to bionanoparticles by measuring the reduction in electrical mobility of a complex particle (particle-protein) that corresponds to an increase in mobility diameter.

Characterization of SAM-functionalized gold nanoparticles by ES-DMA

Citric acid (Figure 13.8) stabilized gold nanoparticles (AuNPs) with diameter in the range 10-60 nm and conjugated AuNPs are analyzed by ES-DMA. This investigation shows that the formation of salt particles on the surface of AuNPs can interfere with the mobility analysis because of the reduction in analyte signals. Since sodium citrate is a non-volatile soluble salt, ES produces two types of droplets. One droplet consists of AuNPs and salt and the other droplet contains only salt. Thus, samples must be cleaned by centrifugation prior to determine the size of bare AuNPs. Figure 13.9 presents the size distribution of AuNPs of distinct diameters and peaks corresponding to salt residues.

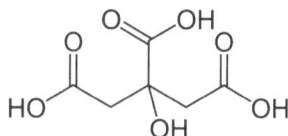

Figure 13.8: Structure of citric acid that provides charge stabilization (as the citrate ion) to AuNPs.

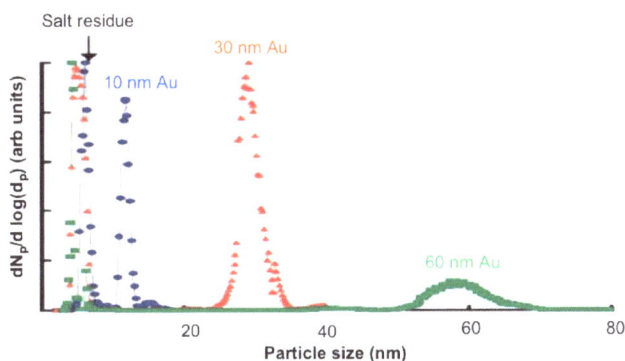

Figure 13.9: Particle size distribution of 10 nm, 30 nm and 60 nm AuNPs after centrifugation cleaning. Adapted from D. Tsai, R. A. Zangmeister, L. F. Pease III, M. J. Tarlov and M. R. Zachariah, Gas-phase ion-mobility characterization of SAM-functionalized Au nanoparticles. *Langmuir*, 2008, 24, 8483. Copyright: American Chemical Society (2015).

The mobility size of bare AuNPs (d_{p0}) can be obtained by using,

$$d_{p0} = \sqrt[3]{d_{p,m}^3 - d_s^3}$$

where $d_{p,m}$ and d_s are mobility sizes of the AuNPs encrusted with salts and the salt NP, respectively. However, the presence of self-assembled monolayer (SAM) produces a difference in electrical mobility between conjugated and bare AuNPs. Hence, the determination of the diameter of AuNPs (salt-free) is critical to distinguish the increment in size after functionalization with SAM. The coating thickness of SAM that corresponds to the change in particle size (ΔL) is calculated by using,

$$\Delta L = d_p - d_{p0}$$

where d_p and d_{p0} are the coated and uncoated particle mobility diameters, respectively.

In addition, the change in particle size can be determined by considering a simple rigid core-shell model that gives theoretical values of ΔL_1 higher than the experimental ones (ΔL). A modified core-shell model is proposed in which a size dependent effect on ΔL_2 is observed for a range of particle sizes.

AuNPs of 10 nm and 60 nm are coated with MUA (Figure 13.10), a charge alkanethiol, and the particle size distributions of bare and coated AuNPs are presented in Figure 13.11. The increment in average particle size is 1.2 ± 0.1 nm for 10 nm AuNPs and 2.0 ± 0.3 nm for 60 nm AuNPs so that ΔL depends on particle size.

Figure 13.10: Structure of 11-mercaptoundecanoic acid (MUA).

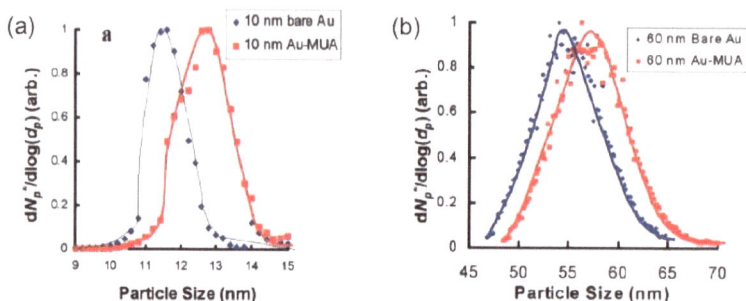

Figure 13.11: Particle size distributions of (a) 10 nm and (b) 60 nm bare versus MUA-coated AuNPs. Adapted from D. Tsai, R. A. Zangmeister, L. F. Pease III, M. J. Tarlov and M. R. Zachariah, Gas-phase ion-mobility characterization of SAM-functionalized Au nanoparticles. *Langmuir*, 2008, 24, 8483. Copyright: American Chemical Society (2015).

Advantages of ES-DMA

- ES-DMA does not need prior information about particle type.
- It characterizes broad particle size range and operates under ambient pressure conditions.
- A few μL or less of sample volume is required and total time of analysis is 2 - 4 min.
- Data interpretation and mobility spectra simple to analyze compared to ES-MS where there are several charge states.

Limitations of ES-DMA

- Analysis requires the following solution conditions: concentrations of a few hundred µg/mL, low ionic strength (<100 mM) and volatile buffers.
- Uncertainty is usually ±0.3 nm from a size range of a few nm to around 100 nm. This is not appropriate to distinguish proteins with slight differences in molecular weight.

Related techniques

A tandem technique is ES-DMA-APM that determines mass of ligands adsorbed to nanoparticles after size selection with DMA. APM is an aerosol particle mass analyzer that measures mass of particles by balancing electrical and centrifugal forces. DMA-APM has been used to analyze the density of carbon nanotubes, the porosity of nanoparticles and the mass and density differences of metal nanoparticles that undergo oxidation.

Bibliography

G. Bacher, W. W. Szymanski, S. T. Kaufman, P. Zollner, D. Blaas, and G. Allmaier. *J. Mass Spectrom.*, 2001, **36**, 1038.

R. C. Flagan, Differential mobility analysis of aerosols: a tutorial. *KONA Powder Part J.*, 2008, **26**, 254.

S. Guha, M. Li, M. J. Tarlov, and M. R. Zachariah, Electrospray-differential mobility analysis of bionanoparticles, *Trends Biotechnol.*, 2012, **30**, 291.

S. Guha, X. Ma, M. J. Tarlov, and M. R. Zachariah, Quantifying ligand adsorption to nanoparticles using tandem differential mobility mass analysis. *Anal. Chem.*, 2012, **84**, 6308.

P. Intra and N. Tippayawong, An overview of differential mobility analyzers for size classification of nanometer-sized aerosol particles. *Songklanakarin J. Sci. Technol.*, 2008, **30**, 243.

S. T. Kaufman, J. W. Skogen, F. D. Dorman, and F. Zarrin, Macromolecule analysis based on electrophoretic mobility in air: globular proteins. *Anal. Chem.*, 1996, **68**, 1895.

E. O. Knutson and K. T. Whitby. Aerosol classification by electric mobility: apparatus, theory, and applications. *J. Aerosol Sci.*, 1975, **6**, 443.

M. Li, S. Guha, R. Zangmeister, M. J. Tarlov, and M. R. Zachariah, Method for determining the absolute number concentration of nanoparticles from electrospray sources. *Langmuir*, 2011, **27**, 14732.

L. F. Pease III, Physical analysis of virus particles using electrospray differential mobility analysis. *Trends Biotechnol.*, 2012, **30**, 216.

D. Tsai, R. A. Zangmeister, L. F. Pease III, M. J. Tarlov and M. R. Zachariah, Gas-phase ion-mobility characterization of SAM-functionalized Au nanoparticles. *Langmuir*, 2008, **24**, 8483

D. Tsai, L. F. Pease III, R. A. Zangmeister, M. J. Tarlov, and M. R. Zachariah, Aggregation kinetics of colloidal particles measured by gas-phase differential mobility analysis. *Langmuir*, 2009, **25**, 140.

Y. Tseng and L. F. Pease III, Electrospray differential mobility analysis for nanoscale medicinal and pharmaceutical applications. *Nanomed-Nanotechnol.*, 2014, **10**, 1591.

Chapter 14: Thermal Conductivity

Juan Velazquez and Andrew R. Barron

Introduction

A catalyst is a "substance that accelerates the rate of chemical reactions without being consumed". Some reactions, such as the hydrodechlorination of trichloroethene (TCE, Figure 14.1),

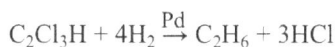

$$C_2Cl_3H + 4H_2 \xrightarrow{Pd} C_2H_6 + 3HCl$$

don't occur spontaneously but can occur in the presence of a catalyst.

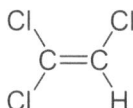

Figure 14.1: Structure of trichloroethene (TCE).

Metal dispersion is a common term within the catalyst industry. The term refers to the amount of metal that is active for a specific reaction. Let's assume a catalyst material has a composition of 1 wt% palladium and 99% alumina (Al_2O_3) (Figure 14.2). Even though the catalyst material has 1 wt% of palladium, not all the palladium is active. The material might be oxidized due to air exposure or some of the material is not exposed to the surface (Figure 14.3), hence it can't participate in the reaction. For this reason, it is important to characterize the material.

Figure 14.2: A sample of commercially available 1 wt% Pd/Al₂O₃.

Figure 14.3: Representation of Pd nanoparticles on Al₂O₃. Some palladium atoms are exposed to the surface, while some other lay below the surface atoms and are not accessible for reaction.

In order for Pd to react as a catalyst for TCE hydrodechlorination, it needs to be in the metallic form. Any oxidized palladium will be inactive. Thus, it is important to determine the oxidation state of the Pd atoms on the surface of the material. This can be accomplished using an experiment called temperature programmed reduction (TPR). Subsequently, the percentage of active palladium can be determined by hydrogen chemisorption. The percentage of active metal is an important parameter when comparing the performance of multiple catalyst. Usually the rate of reaction is normalized by the amount of active catalyst.

Principles of thermal conductivity

Thermal conductivity is the ability of a chemical specie to conduct heat. Each gas has a different thermal conductivity. The units of thermal conductivity in the international system of units are W/m·K. Table 14.1 shows the thermal conductivity of some common gasses.

Gas	Thermal conductivity (W/m·K)
Hydrogen	0.18050
Argon	0.01772
Helium	0.15130
Carbon monoxide	0.02614

Table 14.1: Thermal conductivity values for common gasses.

Thermal conductivity detector

A thermal conductivity detector has four filaments that change resistance according to the thermal conductivity of the gas flowing over it. Two filaments measure the reference gas and the other two measures the sample gas. The detector is isothermal; it will increase or decrease the voltage in each of the resistors in order to maintain a constant temperature. The temperature of the detector is 125 °C. When both the reference and samples gas have the same composition and same flow rate, the resistors are balanced, and the detector will zero the signal. If there is a change in flow rate or in the gas composition the detector will react to maintain the constant temperature. The detector circuitry can be described using a Wheatstone bridge configuration as shown in Figure 14.4. If the gas flowing through the sample has a higher thermal conductivity the filament will cool down, the detector will apply a higher voltage to keep a constant temperature and this will be recorded as a positive signal.

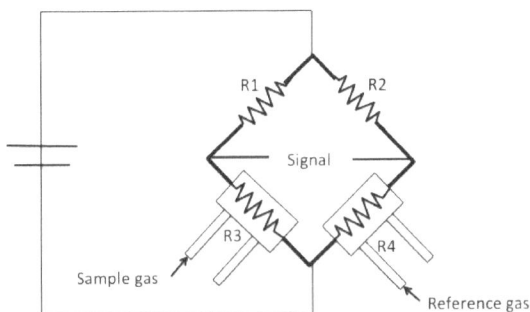

Figure 14.4: A simplified circuit diagram of a thermal conductivity detector.

This detector is part of a typical commercial instrument such as a Micromeritics AutoChem 2920 (Figure 14.5). This instrument is an automated analyzer with the ability to perform chemical adsorption and temperature-programmed reactions on a catalyst, catalyst support, or other materials.

Figure 14.5: A photograph of a Micromeritics AutoChem 2920.

Temperature programmed reduction (TPR)

TPR will determine the number of reducible species on a catalyst and will tell at what temperature each of these species was reduced. For example palladium is ordinarily found as Pd(0) or Pd(II), i.e., oxidation states 0 and +2. Pd(II) can be reduced at very low temperatures (5 - 10 °C) to Pd(0) following,

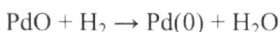

$$PdO + H_2 \rightarrow Pd(0) + H_2O$$

A 128.9 mg 1wt% Pd/Al$_2$O$_3$ samples is used for the experiment (Figure 14.6). Since we want to study the oxidation state of the commercial catalyst, no pre-treatment needs to be executed to the sample. A 10% hydrogen-argon mixture is used as analysis and reference gas. Argon has a low thermal conductivity and hydrogen has a much higher thermal conductivity. All gases will flow at 50 cm^3/min. The TPR experiment will start at an initial temperature of 200 K, temperature ramp 10 K/min, and final temperature of 400 K. The H$_2$/Ar mixture is flowed through the sample, and past the detector in the analysis port. While in the reference port the mixture doesn't become in contact with the sample. When the analysis gas starts flowing over the sample, a baseline reading is established by the detector. The baseline is established at the initial temperature to ensure there is no reduction. While this gas is flowing, the temperature of the sample is increased linearly with time and the consumption of hydrogen is recorded. Hydrogen atoms react with oxygen atoms to form H$_2$O.

Figure 14.6: A sample of Pd/Al$_2$O$_3$ in a typical sample holder.

Water molecules are removed from the gas stream using a cold trap. As a result, the amount of hydrogen in the argon/hydrogen gas mixture decreases and the thermal conductivity of the mixture also decrease. The change is compared to the reference gas and yields to a hydrogen uptake volume. Figure 14.7 is a typical TPR profile for PdO.

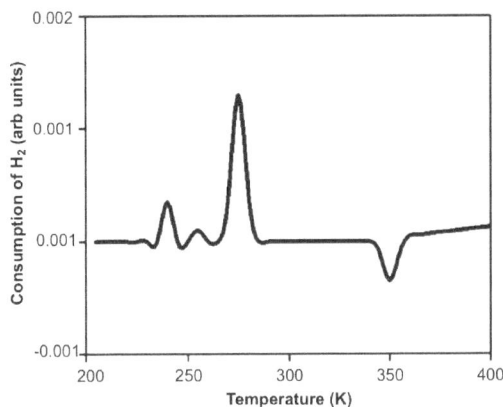

Figure 14.7: A typical TPR profile of PdO. Adapted from R. Zhang, J. A. Schwarz, A. Datye, and J. P. Baltrus, The effect of second-phase oxides on the catalytic properties of dispersed metals: palladium supported on 12% WO₃/Al₂O₃. *J. Catal.*, 1992, 138, 55. Copyright: Elsevier (1992).

Pulse chemisorption

Once the catalyst (1 wt% Pd/Al₂O₃) has been completely reduced, the user will be able to determine how much palladium is active. A pulse chemisorption experiment will determine active surface area, percentage of metal dispersion and particle size. Pulses of hydrogen will be introduced to the sample tube in order to interact with the sample. In each pulse hydrogen will undergo a dissociative adsorption on to palladium active sites until all palladium atoms have reacted. After all active sites have reacted, the hydrogen pulses emerge unchanged from the sample tube. The amount of hydrogen chemisorbed is calculated as the total amount of hydrogen injected minus the total amount eluted from the system.

Data collection for hydrogen pulse chemisorption

The sample from previous experiment (TPR) will be used for this experiment. Ultra-high-purity argon will be used to purge the sample at a flow rate of 40 cm^3/min. The sample will be heated to 200 °C in order to remove all chemi-sorbed hydrogen atoms from the Pd(0) surface. The sample is cooled down to 40 °C. Argon will be used as carrier gas at a flow of 40 cm^3/min. Filaments temperature will be 175 °C and the detector temperature will be 110 °C. The injection loop has a volume of 0.03610 cm^3 @ STP. As shown in Figure 14.8, hydrogen pulses will be injected into the flow stream, carried by argon to be-come in contact and react with the sample. It should be noted that the first pulse of hydrogen was almost completely adsorbed by the sample. The second and third pulses show how the samples is been saturated. The positive value of the TCD detector is consistent with our assumptions. Since hydrogen has a higher thermal conductivity than argon, as it flows through the detector it will tend to cool down the filaments, the detector will then apply a positive voltage to the filaments in order to maintain a constant temperature.

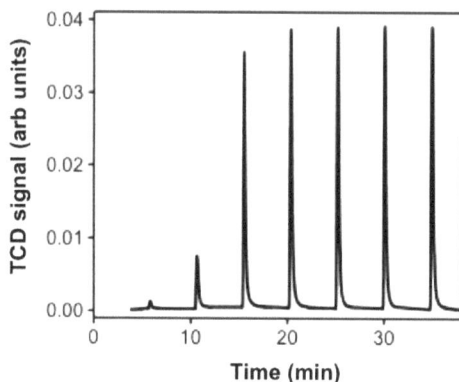

Figure 14.8: A typical hydrogen pulse chemisorption profile of 1 wt% Pd/Al$_2$O$_3$.

Pulse chemisorption data analysis

Table 14.2 shows the integration of the peaks from Figure 14.8. This integra-tion is performed by an automated software provided with the instrument. It should be noted that the first pulse was completely consumed by the sample, the pulse was injected between time 0 and 5 minutes. From Figure 14.8 we observe that during the first four pulses, hydrogen is consumed by the sample. After the fourth pulse, it appears the sample is not consuming hydrogen. The experiment continues for a total of seven pulses, at this point the software determines that no consumption is occurring and stops the experiment. Pulse eight is denominated the "saturation peak", meaning the pulse at which no hydrogen was consumed.

Pulse n	Area
1	0
2	0.000471772
3	0.00247767
4	0.009846683
5	0.010348201
6	0.010030243
7	0.009967717
8	0.010580979

Table 14.2: Hydrogen pulse chemisorption data.

Using,

$$\Delta Area_n = Area_{saturation} - Area_n$$

the change in area ($\Delta area_n$) is calculated for each peak pulse area ($area_n$)and compared to that of the saturation pulse area ($area_{saturation} = 0.010580979$). Each of these changes in area is proportional to an amount of hydrogen consumed by the sample in each pulse. Table 14.3 shows the calculated change in area.

Pulse n	$Area_n$	$\Delta Area_n$
1	0	0.010580979
2	0.000471772	0.0105338018
3	0.00247767	0.008103309
4	0.009846683	0.000734296
5	0.010348201	0.000232778
6	0.010030243	0.000550736
7	0.009967717	0.000613262
8	0.010580979	0

Table 14.3: Hydrogen pulse chemisorption data with ΔArea.

The $\Delta area_n$ values are then converted into hydrogen gas consumption using,

$$V_{adsorbed} = \frac{\Delta Area_n \times F_c}{SW}$$

where F_c is the area-to-volume conversion factor for hydrogen and SW is the weight of the sample. F_c is equal to 2.6465 cm^3/peak area. Table 14.4 shows the results of the volume adsorbed and the cumulative volume adsorbed. Using the data on Table 14.4, a series of calculations can now be performed in order to have a better understanding of our catalyst properties.

Pulse n	Area$_n$	ΔArea$_n$	V$_{adsorbed}$ (cm^3/g STP)	Cumulative quantity (cm^3/g STP)
1	0	0.010580979	0.2800256	0.2800256
2	0.000471772	0.0105338018	0.2787771	0.5588027
3	0.00247767	0.008103309	0.2144541	0.7732567
4	0.009846683	0.000734296	0.0194331	0.7926899
5	0.010348201	0.000232778	0.0061605	0.7988504
6	0.010030243	0.000550736	0.0145752	0.8134256
7	0.009967717	0.000613262	0.0162300	0.8296556
8	0.010580979	0	0.0000000	0.8296556

Table 14.4: Hydrogen pulse chemisorption data including the volume adsorbed per pulse and the cumulative volume adsorbed.

Gram molecular weight

Gram molecular weight is the weighted average of the number of moles of each active metal in the catalyst. Since this is a monometallic catalyst, the gram molecular weight is equal to the molecular weight of palladium (106.42 [g/mol]). The GMC$_{Calc}$ is calculated using

$$GMW_{Calc} = \frac{1}{\left(\frac{F_1}{W_{atomic1}}\right) + \left(\frac{F_2}{W_{atomic2}}\right) + \dots + \left(\frac{F_N}{W_{atomicN}}\right)}$$

where F is the fraction of sample weight for metal N and $W_{atomicN}$ is the gram molecular weight of metal N (g/g-mole). The calculation for this experiment is shown in

$$GMW_{Calc} = \frac{1}{\left(\frac{F_1}{W_{atomicPd}}\right)} = \frac{W_{atomicPd}}{F_1} = \frac{106.42\frac{g}{g\text{-mole}}}{1} = 106.42\frac{g}{g\text{-mole}}$$

Metal dispersion

The metal dispersion is calculated using,

$$GMW_{Calc} = \cfrac{1}{\left(\cfrac{F_1}{W_{atomic1}}\right) + \left(\cfrac{F_2}{W_{atomic2}}\right) + \dots + \left(\cfrac{F_N}{W_{atomicN}}\right)}$$

where PD is the percent metal dispersion, V_s is the volume adsorbed (cm^3 at STP), SF$_{Calc}$ is the calculated stoichiometry factor (equal to 2 for a palladium-hydrogen system), SW is the sample weight and GMW$_{Calc}$ is the calculated gram molecular weight of the sample [g/g-mole]. Therefore, in,

$$GMW_{Calc} = \cfrac{1}{\left(\cfrac{F_1}{W_{atomicPd}}\right)} = \cfrac{W_{atomicPd}}{F_1} = \cfrac{106.42\,\frac{g}{g\text{-mole}}}{1} = 106.42\,\frac{g}{g\text{-mole}}$$

we obtain a metal dispersion of 6.03%.

Metallic surface area per gram of metal

The metallic surface area per gram of metal is calculated using,

$$SA_{Metallic} = \left(\cfrac{V_S}{SW_{Metal} \times 22414}\right) \times (SF_{Calc}) \times \left(6.022 \times 10^{23}\right) \times SA_{Pd}$$

where SA$_{Metallic}$ is the metallic surface area (m^2/g of metal), SW$_{Metal}$ is the active metal weight, SF$_{Calc}$ is the calculated stoichiometric factor and SA$_{Pd}$ is the cross-sectional area of one palladium atom (nm^2). Thus, we obtain a metallic surface area of 2420.99 m^2/g-metal,

$$SA_{Metallic} = \left(\cfrac{0.8296556\,[cm^3]}{0.001289\,[g_{metal}] \times 22414\,\left[\frac{cm^3}{mol}\right]}\right) \times (2) \times \left(6.022 \times 10^{23}\,\left[\frac{atoms}{mol}\right]\right) \times 0.07\,\left[\frac{nm^2}{atom}\right] = 2420.99\,\left[\frac{m^2}{g\text{-metal}}\right]$$

Active particle size

The active particle size is estimated using,

$$APS = \cfrac{6}{D_{Calc} \times \left(\cfrac{W_s}{GMW_{Calc}}\right) \times \left(6.022 \times 10^{23}\right) \times SA_{Metallic}}$$

where D$_{Calc}$ is palladium metal density (g/cm^3), SW$_{Metal}$ is the active metal weight, GMW$_{Calc}$ is the calculated gram molecular weight (g/g-mole), and

SA_{Pd} is the cross-sectional area of one palladium atom (nm^2). Thus, we obtain an optical particle size of 2.88 nm.

$$APS = \frac{600}{\left(1.202 \times 10^{-20}\left[\frac{g_{Pd}}{nm^3}\right]\right) \times \left(\frac{0.001289[g]}{106.42\left[\frac{g_{Pd}}{mol}\right]}\right) \times \left(6.022 \times 10^{23}\left[\frac{atoms}{mol}\right]\right) \times \left(2420.99\left[\frac{m^2}{g_{Pd}}\right]\right)} = 2.88nm$$

In a commercial instrument, a summary report (Table 14.5) will be provided which summarizes the properties of our catalytic material. All the equations used during this example were extracted from the AutoChem 2920-User's Manual.

Properties	Value
Palladium atomic weight	106.4 g/mol
Atomic cross-sectional area	0.0787 nm²
Metal density	12.02 g/cm³
Palladium loading	1 wt%
Metal dispersion	6.03%
Metallic surface area	2420.99 m²/g-metal
Active particle diameter (hemisphere)	2.88 nm

Table 14.5: Summary report provided by Micromeritics AuthoChem 2920.

Bibliography

A. J. Canty, Development of organopalladium(IV) chemistry: fundamental aspects and systems for studies of mechanism in organometallic chemistry and catalysis. *Acc. Chem. Res.*, 1992, **25**, 83.

H. S. Fogler, *Elements of Chemical Reaction Engineering*, Prentice-Hall, New York (1992).

Micromeritics Instrument Corporation, *AutoChem 2920 - Automated catalyst characterization system – Operators Manual*, V4.01 (2009).

M. O. Nutt, K. N. Heck, P. Alvarez, and M. S. Wong, Improved Pd-on-Au bimetallic nanoparticle catalysts for aqueous-phase trichloroethene hydrodechlorination. *Appl. Catal.*, 2006, **69**, 115.

M. O. Nutt, J. B. Hughes, and M. S. Wong, Designing Pd-on-Au bimetallic nanoparticle catalysts for trichloroethene hydrodechlorination. *Environ. Sci. Technol.*, 2005, **39**, 1346.

P. A. Webb and C. Orr, *Analytical Methods in Fine Particle Technology*, Micromeritics Instrument Corp (1997).

R. Zhang, J. A. Schwarz, A. Datye, and J. P. Baltrus, The effect of second-phase oxides on the catalytic properties of dispersed metals: palladium supported on 12% WO_3/Al_2O_3. *J. Catal.*, 1992, **138**, 55.

Chapter 15: Crystal Microbalance

Daniel Garcia-Rojas and Andrew R. Barron

Introduction

The working principle of a quartz crystal microbalance with dissipation (QCM-D) module is the utilization of the resonance properties of some piezoelectric of materials. A piezoelectric material is a material that exhibits an electrical field when a mechanical strain is applied. This phenomenon is also observed in the contrary where an applied electrical field produce a mechanical strain in the material. The material used is α-SiO_2 that produces a very stable and constant frequency. The direction and magnitude of the mechanical strain is directly dependent of the direction of the applied electrical field and the inherent physical properties of the crystal.

A special crystal cut is used, called AT-cut, which is obtain as wafers of the crystal of about 0.1 to 0.3 mm in width and 1 cm in diameter. The AT-cut is obtained when the wafer is cut at 35.25° of the main crystallographic axis of SiO_2. This special cut allows only one vibration mode, the shear mode, to be accessed and thus exploited for analytical purposes. When an electrical field is applied to the crystal wafer via metal electrodes, that are vapor-deposited in the surface, a mechanical shear is produced and maintained as long as the electrical field is applied. Since this electric field can be controlled by opening and closing an electrical circuit, a resonance within the crystal is formed (Figure 15.1).

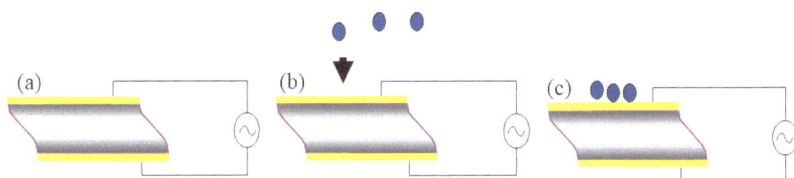

Figure 15.1: Schematic representation of the piezoelectric material: (a) a baseline is obtained by running the sensor without any flow or sample; (b) sample is starting to flow into the sensor; (c) sample deposited in the sensor change the frequency.

Since the frequency of the resonance is dependent of the characteristics of the crystal, an increase of mass, for example when the sample is loaded into the sensor would change the frequency change. This relation was obtained by Sauerbrey in 1959,

$$\Delta m = -C(^1/_n)\Delta f$$

where Δm (ng.cm^{-2}) is the areal mass, C (17.7 ngcm^{-2}Hz^{-1}) is the vibrational constant (shear, effective area, etc.), n in Hz is the resonant overtone, and Δf is the change in frequency. The dependence of the change in the frequency can be related directly to the change in mass deposited in the sensor only when three conditions are met and assumed:

- The mass deposited is small compared to the mass of the sensor.
- It is rigid enough so that it vibrates with the sensor and does not suffer deformation.
- The mass is evenly distributed among the surface of the sensor.

An important incorporation in recent equipment is the use of the dissipation factor. The inclusion of the dissipation faster takes into account the weakening of the frequency as it travels along the newly deposited mass. In a rigid layer the frequency is usually constant and travels through the newly formed mass without interruption, thus, the dissipation is not important. On the other hand, when the deposited material has a soft consistency the dissipation of the frequency is increased. This effect can be monitored and related directly to the nature of the mass deposited.

The applications of the QCM-D ranges from the deposition of nanoparticles into a surface, from the interaction of proteins within certain substrates. It can also monitor the amount of bacterial products when feed with different molecules, as the flexibility of the sensors into what can be deposited in them include nanoparticle, special functionalization or even cell and bacteria.

Experimental planning

In order to use QCM-D for studying the interaction of nanoparticles with a specific surface several steps must be followed. For demonstration purposes the following procedure will describe the use of a Q-Sense E4 with autosampler from Biolin Scientific. A summary is shown below as a quick guide to follow, but further details will be explained:

- Surface election and cleaning according with the manufacturer recommendations.

- Sample preparation including having the correct dilutions and enough sample for the running experiment.
- Equipment cleaning and set up of the correct parameters for the experiment.
- Data acquisition.
- Data interpretation.

Surface election

The decision of what surface of the sensor to use is the most important decision to make for each study. Biolin has a large library of available coatings ranging from different compositions of pure elements and oxides (Figure 15.2) to specific binding proteins. It is important to take into account the different chemistries of the sensors and the results we are looking for. For example, studying a protein with high sulfur content on a gold sensor can lead to a false deposition results, as gold and sulfur have a high affinity to form bonds. For the purpose of this example, a gold coated sensor will be used in the remainder of the discussion.

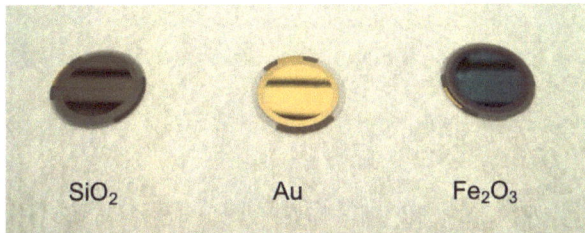

Figure 15.2: Photograph of 1 cm in diameter silica (SiO_2), gold (Au), and iron oxide (Fe_2O_3) coated sensors.

Sensor cleaning

Since QCM-D relies on the amount of mass that is deposited into the surface of the sensor, a thorough cleaning is needed to ensure there is no contaminants on the surface that can lead to errors in the measurement. The procedure the manufacturer established to clean a gold sensor is as follows:
1. Put the sensor in the UV/ozone chamber for 10 minutes
2. Prepare 10 mL of a 5:1:1 solution of hydrogen peroxide:ammonia:water
3. Submerge in this solution at 75 °C for 5 minutes.
4. Rinse with copious amount of milliQ water.

5. Dry with inert gas.
6. Put the sensor in the UV/ozone chamber for 10 minutes as shown in Figure 15.3.

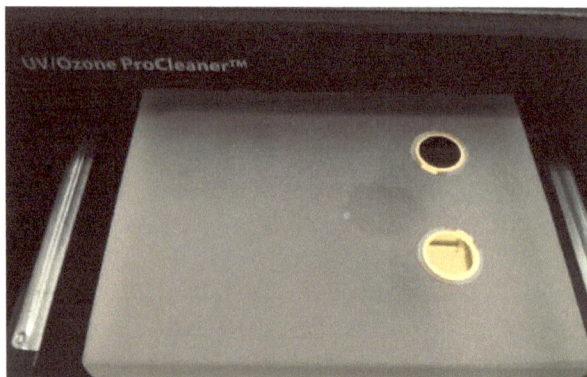

Figure 15.3: Gold sensors in loader of the UV/ozone chamber in the final step of the cleaning process.

Once the sensors are clean, extreme caution should be taken to avoid contamination of the surface. The sensors can be loaded in the flow chamber of the equipment making sure that the T-mark of the sensor matches the T mark of the chamber in order to make sure the electrodes are in constant contact. The correct position is shown in Figure 15.4.

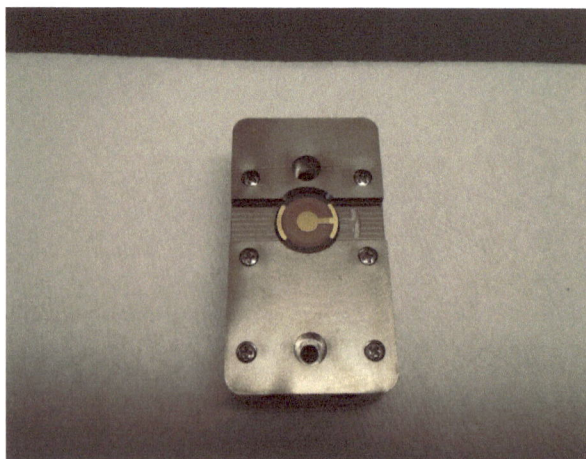

Figure 15.4: Correct position of the sensor in the chamber.

Sample preparation

As the top range of mass that can be detected is merely micrograms, solutions must be prepared accordingly. For a typical run, a buffer solution is needed in which the deposition will be studied as well as, the sample itself and a solution of 2% of sodium dodecylsulfate (SDS, Figure 15.5). For this example, we will be using nanoparticles of magnetic iron oxide (nMag) coated with PAMS, and as a buffer 8% NaCl in DI water.

- For the nanoparticles sample it is necessary to make sure the final concentration of the nanoparticles will not exceed 1 mM.
- For the buffer solution, it is enough to dissolve 8 g of NaCl in DI water.
- For the SDS solution, 2 g of SDS should be dissolved very slowly in approximate 200 mL of DI water, then 100 mL aliquots of DI water are added until the volume is 1 L. This is in order to avoid the formation of bubbles and foam in the solution.

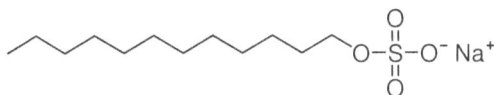

Figure 15.5: Structure of sodium dodecylsulfate [$CH_3(CH_2)_{10}CH_2OSO_3Na$, SDS].

Instrument preparation

Due to the sensitivity of the equipment, it is important to rinse and clean the tubing before loading any sample or performing any experiments. To rinse the tubing and the chambers, use a solution of 2% of SDS. For this purpose, a cycle in the autosampler equipment is program with the steps shown in Table 15.1.

Step	Duration (min)	Speed (µL/min)	Volume (mL)
DI water (1:2)	10	100	1
SDS (1:1)	20	300	6
DI water (1:2)	10	100	1

Table 15.1: Summary of cleaning processes.

Once the equipment is cleaned, it is ready to perform an experiment, a second program in the autosampler is loaded with the parameters shown in Table 15.2.

Step	Duration (min)	Speed (µL/min)	Volume (mL)
Buffer (1:3)	7	100	0.7
Nanoparticles (1:4)	30	100	3.0

Table 15.2: Experimental set-up.

The purpose of flowing the buffer in the beginning is to provide a background signal to take into account when running the samples. Usually a small quantity of the sample is loaded into the sensor at a very slow flow rate in order to let the deposition take place.

Data acquisition

Example data obtained with the above parameters is shown in Figure 15.6. The blue squares depict the change in the frequency. As the experiment continues, the frequency decreases as more mass is deposited. On the other hand, shown as the red squares, the dissipation increases, describing the increase of both the height and certain loss of the rigidity in the layer from the top of the sensor. To illustrate the different steps of the experiment, each section has been color-coded. The blue part of the data obtained corresponds to the flow of the buffer, while the yellow part corresponds to the deposition equilibrium of the nanoparticles onto the gold surface. After certain length of time equilibrium is reached and there is no further change. Once equilibrium indicates no change for about five minutes, it is safe to say the deposition will not change.

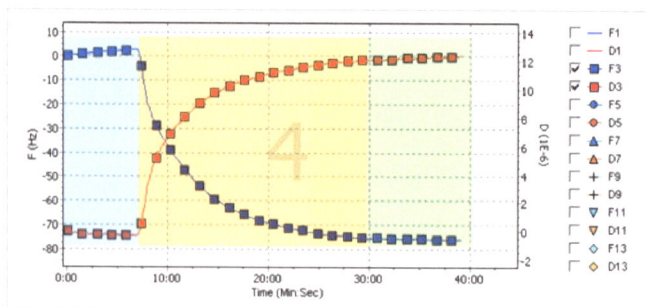

Figure 15.6: Data of deposition of nMag in a gold surface.

Instrument clean-up

As a measure preventive care for the equipment, the same cleaning procedure should be followed as what was done before loading the sample. Use of a 2% solution of SDS helps to ensure the equipment remains as clean as possible.

Data modeling

Once the data has been obtained, QTools (software that is available in the software suit of the equipment) can be used to convert the change in the frequency to areal mass, via the Sauerbrey equation. The correspondent graph of areal mass is shown in Figure 15.7. From this graph we can observe how the mass is increasing as the nMag is deposited in the surface of the sensor. The blue section again illustrates the part of the experiment where only buffer was been flown to the chamber. The yellow part illustrates the deposition, while the green part shows no change in the mass after a period of time, which indicates the deposition is finished. The conversion from areal mass to mass is a simple process, as gold sensors come with a definite area of 1 cm^2, but a more accurate measure should be taken when using functionalized sensors.

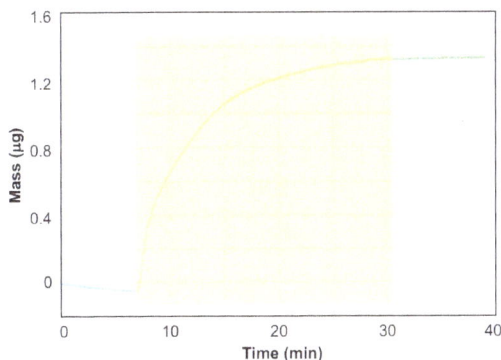

Figure 15.7: Mass of deposition of nMag into gold surface.

It is important to take into account the limitations of the Saubery equation, because the equation accounts for a uniform layer on top of the surface of the sensor. Deviations due to clusters of material deposited in one place or the formation of partial multilayers in the sensor cannot be calculated through this model. Further characterization of the surface should be done to have a more accurate model of the phenomena.

Bibliography

Biolin Scientific, *Cleaning and Immobilization Protocols* (2004).

F. Hook, *Development of a Novel QCM Technique for Protein Adsorption Studies*, Chalmers University (1997).

C. Ziez, *Theoretical and Experimental Analysis on Nanoparticle-Nanoparticle and Nanoparticle-Surface Interactions and their Role in Defining their Stability and Mobility*, Rice University (2013).